Information Ecologies

Information Ecologies
Using Technology with Heart

Bonnie A. Nardi and Vicki L. O'Day

The MIT Press
Cambridge, Massachusetts
London, England

© 1999 Massachusetts Institute of Technology

All rights reserved. No part of this book may be reproduced in any form by any electronic or mechanical means (including photocopying, recording, or information storage and retrieval) without permission in writing from the publisher.

This book was set in Sabon by Wellington Graphics.
Printed and bound in the United States of America.

Library of Congress Cataloging-in-Publication Data

Nardi, Bonnie A.
  Information ecologies : using technology with heart /
Bonnie A. Nardi and Vicki L. O'Day.
    p. cm.
  Includes bibliographical references and index.
  ISBN 0-262-14066-7 (hc : alk. paper)
  1. Technology—Social aspects. I. O'Day, Vicki. II. Title.
T14.5.N344 1998
303.48′3—dc21   98-29318
                                                                         CIP

To our husbands, Chris Darrouzet and Eduardo Pelegrí-Llopart

Contents

Preface

One of the most important human stories of the twentieth century is the impact of technology on the way we live, die, work, and play. This will continue into the twenty-first century. Usually discussions of technology are either blissfully pro or darkly con. Most of the time, people do not discuss technology at all. They simply let it wash over them, adapting as best they can. This book is an attempt to engender a public conversation that will be more balanced and nuanced, to develop a critical stance that is less passive and unreflectively accepting.

There are reasons to be concerned about the impacts of technology—the rapid pace of technological change challenges our ability to keep up, human skill and judgment at work are lost to automation, and standards of mechanical efficiency are used as benchmarks for human performance.

We see ourselves as critical friends of technology. We believe we can find ways to enjoy the fruits of technology without being diminished by it. It is possible to use technology with pleasure and grace if we make thoughtful decisions in the context of our "local habitations," to borrow Shakespeare's phrase. By this we mean settings in which we as individuals have an active role, a unique and valuable local perspective, and a say in what happens. For most of us, this means our workplaces, schools, homes, libraries, hospitals, community centers, churches, clubs, and civic organizations. For some of us, it means a wider sphere of influence. All of us have local habitations in which we can reflect on appropriate uses of technology in light of our local practices, goals, and values.

We call these local habitations "information ecologies," since they have much in common with biological ecologies, as we will discuss. Because the goal of this book is to change the way people look at technology in their own settings, we adopted a metaphor that emphasizes local connections and offers scope for diverse reflections and analyses. We believe that we have leverage to affect technological change by acting in spheres where we have knowledge and authority—our own information ecologies. A key to thoughtful action is to ask many more "know-why" questions than we typically do. Being efficient, productive, proactive people, we often jump to the "know-how" questions, which are considerably easier to answer. In this book we talk about practical ways to have more "know-why" conversations, to dig deeper, and reflect more about the effects of the ways we use technology.

The phrase "local habitations" helps us understand settings of technology use in a new and useful way. Fritz Lang's beautiful film *Metropolis* is another source of insight for us. *Metropolis* engages some of our collective fears about our society's dependence on technological invention. The film presents a view of technology as a seductive, untamable force that undermines our humanity. In 1926, Lang sensed the way technology would keep apart heart and mind, the way people would heedlessly focus on technical development for its own sake while evading the social questions of what purpose technology serves in human life.

Rotwang, the unforgettable mad scientist in *Metropolis*, created the ultimate robot, a creature possessed of full human intelligence. Lang recognized the deep love that goes into technical creation—the robot was created in the image of Rotwang's beloved dead mistress. Rotwang refused to consider how such a robot might be used for evil, and indeed, heartless forces of capitalism harness the powers of Maria, the robot. It is important that we understand the message Lang was sending us: we love our technologies and we are endlessly technically creative, but our creations can betray us. Rotwang was too entranced with his invention to consider the possible human consequences. J. Robert Oppenheimer, in a similar vein, said of the development of the hydrogen bomb that the mere fact of the possibility of creating the bomb "was technically so sweet that you could not argue about that."[1]

We believe that we can and should argue about how technology is created and used. Lang suggested in *Metropolis* that technical sweetness is not enough. Technology development and use must be mediated by the human heart.

In this book, we discuss what it could mean to use technology with heart. We give examples from our research studies, to show how people can use technology fruitfully by engaging their own values and commitments. We examine the groundbreaking analyses of scholars such as Jacques Ellul and Langdon Winner, who have deepened our understanding through their provocative looks at the social implications of technology. We hope that these examples and ideas will help you see new avenues of participation and engagement with technology in your own local settings.

Acknowledgments

We would like to thank Frances Jacobson and Jeff Johnson for their supportive and helpful early readings of the manuscript. Steve Cisler offered many pointers to key literature and Web sites that informed our work. Librarians at the Apple Computer and Xerox PARC libraries greatly aided the research for the book. Special thanks to Ed Valauskas, who liberally dispensed moral and practical support and suggested the film *Metropolis* to us. Portions of chapter seven first appeared in *Libri,* and we thank the editors for their generous permission to use the material here. Many thanks to our editor Bob Prior and the helpful staff at the MIT Press. Several anonymous reviewers sent us valuable comments.

I would like to acknowledge my former employer, Apple Computer, where I worked for most of the writing of the book, as well as my new home at AT&T Labs-Research, in the Menlo Park, California, office. My children, Anthony, Christopher, and Jeanette, are the motivating force behind a desire to approach technology more reflectively. Thanks to them for their hugs and humor. Michelle Gantt, Allan Kuchinsky, Robert Leichner, Brian Reilly, Heinrich Schwarz, and Steve Whittaker were valued colleagues in some of the research reported here. Trish Lynch and Cliff Herlth supported the research at Lincoln High in every possible way, and inspired me with their dedication. I thank Monica Ertel, and all the people at the Apple Library for a transforming experience as well as the continuing contact with the library community that I have been privileged to enjoy. I can't imagine how I would have met so many extraordinary librarians, brain surgeons, teenagers, teachers, and spreadsheet users if I weren't an anthropologist, and I owe a special debt of gratitude to the informants in my studies who gave freely of their time and thoughts. I thank my husband Chris Darrouzet with whom I have shared a long and loving partnership.

B. N.

I thank Xerox PARC, SRI International, and Hewlett-Packard for supporting my migrations between software design and ethnography. Pueblo was a particularly transformative research project, and I am grateful to all the members of Pueblo for inspiration and companionship. Thanks to lead investigators Daniel Bobrow and Vijay Saraswat and other group members Bjorn Carlson, Vineet Gupta, and Mark Shirley at Xerox PARC; Billie Hughes, Rose Pfefferbaum, and Jim Walters at Phoenix College; Bill Lightfoot, Anne Mattison, Paula Melton, Cyndy Olson, Donna Spano, Jo Talazus, Cynde Welbes, and Curt Whowell at Longview Elementary School; the children, grays, and many others who make Pueblo such a special place; and most of all, Kim Bobrow, a very wise and kindly wizard. I have enjoyed and learned from many provocative conversations about network communities with my SeniorNet research partners: Annette Adler, Mimi Ito, Charlotte Linde, and Beth Mynatt. I also thank my colleagues in library research, Robin Jeffries and Andreas Paepcke, and the Hewlett-Packard librarians and library clients who informed my library studies. I appreciate the ongoing interest and help of Alice Wilder Hall, Maia Pindar, and other librarians at Xerox PARC. Special thanks are due to my mother, Margaret O'Day, whose splendid offer of last-minute child care allowed me to finish this book on schedule. Lia Adams, Kate Finn, Liz Lada, and Kathie Nichols provided encouragement, baby-sitting, and tea. I am grateful to my children, Emma and Patrick, for leading me in many new directions. I thank my husband, Eduardo, for always giving me his loving support, even while he was living on Internet time.

V. O.

I

Information Ecologies: Concepts and Reflections

1
Rotwang the Inventor

Fritz Lang's classic film *Metropolis* was released in Berlin in January 1927. Computers existed only as primitive Hollerith cards, but electricity, automobiles, airplanes, and telephones had entered the scene by the time Lang created the first science fiction movie. Lang's film presents an extraordinary and prescient vision of the seductive appeal and sheer beauty of technology, along with the potentially dehumanizing effects on those who are slaves to its operation and those who would claim to be its masters. Though the movie's plot and characters are idealized and simplistic, the complex and beautifully composed images of Lang's future world are unforgettable. The film's themes form a backdrop for our reflections on people and technology.[1]

Here is the city of Metropolis in the twenty-first century. Above ground, the city's immense buildings create breathtaking patterns of light, shadow, and geometric form. Crowded roadways are suspended high above the ground, crisscrossing the vast spaces between buildings. The cityscape is visually stunning, built by human hands on a scale that transcends ordinary human activities.

Whose hands? Not those of the masters of Metropolis—the masters provide only the minds that direct the city from above. The hands belong to the people who live and work in the depths below ground, slaves to the machines that run the city. We first see the workers through the eyes of the film's hero, the son of one of the masters, who is curious to see what life is like for his brothers and sisters underground.

He descends and enters a huge space swirling with steam from the machines. He sees a wall many stories high covered with an array of dials, wheels, gears, and levers, with platforms where individual workers stand, desperately focused on the controls facing them. From afar, the workers appear to be engaged in a compelling mechanical dance, as if each is a moving part within a single gigantic machine. The human-machine interfaces are just beyond the size of the human body. A worker must extend to the limits of his reach to carry out his tasks, stretching, crouching, and twisting in jerky reaction to the blinking lights and moving gauges before him. From a distance there appears to be a collaborative harmony in the workers' choreographed motions. But closer up it is clear that each worker is alone, unable to glance at his companions working nearby because of the incessant demands of the machines.

Metropolis,

A worker's exhaustion leads to a fatal explosion, witnessed by the impressionable visitor from above. "Such accidents are unavoidable," the Master tells his son—but only if you rely on human workers. Enter Rotwang the inventor, a scientist who lives apart from the society of the masters but places himself and his talents at their service. Rotwang announces to the Master that he has made a machine in the image of Man, though interestingly, the gleaming robot is certainly female. Rotwang only needs a human model to imbue his creation with a fully lifelike appearance, which will later occur in a Frankenstein-like transfer of energy that is the first full-screen example of morphing.

The Master finds the perfect candidate in Maria, a lovely, pure daughter of the lower class who preaches hope to the groups of tired workers who assemble secretly in the catacombs to hear her. Someday, she tells them, a mediator will come to bridge the distance between mind and hands with heart. The Master now sees a way of killing two birds with one stone. He asks Rotwang to capture Maria, form the robot in her image, and send it below to impersonate her and incite violence among the workers. Their self-destruction will leave the way clear to replace

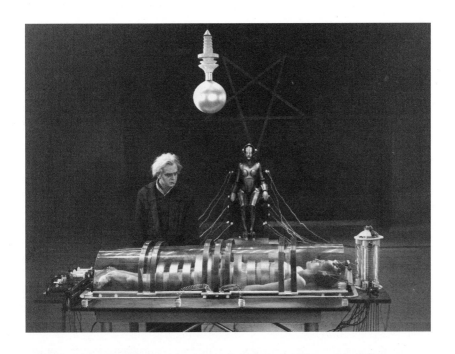

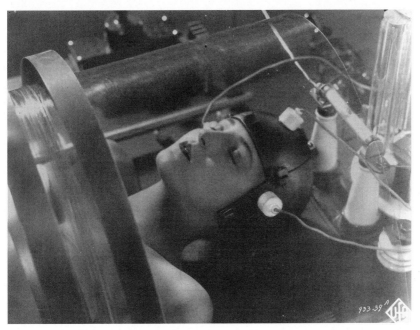

vulnerable human workers with robots who will never grow tired or make mistakes.

Rotwang, an eccentric, distant character, is disturbingly indifferent to the uses of his inventions and the consequences of his actions. He agrees to the Master's request and pursues Maria in the catacombs in a terrifying scene of indirect but menacing violence. Rotwang's only weapon is light—a flashlight that inexorably follows Maria as she dashes from one dark corner to another. She is pinned by the light and carried up to Rotwang's laboratory. Light is used continually throughout the film to represent technology, from the elegant sculptures of light on the Master's desk to the dangerous rings of electrical current that animate the robot.

The robot Maria is astonishingly sexy. She acknowledges her instructions with a slow, heavy wink, and slinks down to begin her work of starting a riot. Lang's layers of meaning here are mind-boggling—a robot is sent to preach the destruction of technology, using human sexuality as a persuasive force.

As the workers begin their attack on the machines, Lang gives us yet another glancing reference to the pervasive use of technology—the Master watches the chaos over a video monitoring system and holds a video conference with his panicking foreman. (In moments like these it is hard to remember that this film was completed in 1926.) As the machinery fails, the walls of an underground reservoir break and the waters begin to rise, spreading silently across the workers' own city in a horrifying and claustrophobic sequence. Disastrous forces of nature are now joined to those of the machines gone awry. The city above is still beautiful, even as it begins to crack and fall.

As the Master's son is rejoined with the real Maria, the workers understand that they have destroyed their own homes and risked their children's lives. They burn the robot at the stake as a witch. Rotwang, maddened by the loss of his invention, struggles with the hero and falls from the roof of a cathedral.

The Master's son entreats his father to reach out and clasp the hand of his own foreman, suggesting a new future in which the minds that plan and the hands that work do not live in separate worlds, but are mediated by the human heart.

2

Framing Conversations about Technology

The seventy-year-old film *Metropolis* is a reminder that our current questions and concerns about technology have a long history. Many of the particular technologies we experience now are fairly new—voicemail, cellular phones, the Internet, and many more. But the challenge of responding well to technological change goes back at least to the invention of the earliest machines.

There is no question about the imaginative appeal of technology, not just in the cityscape of Metropolis but in our own world today. With the help of technology, we can understand genetic structure, take pictures of stars being born, and perform in utero surgery to save the life of an unborn baby. These are accomplishments that give us a sense of wonder and appreciation for human inventiveness. They celebrate our abilities and extend our connections with the natural world.

When we adopt new technologies, we face uncertainty about how our quality of life may change. The development of new technology affects the nature of work, school, family life, commerce, politics, and war. We might expect that anything with such profound influence on the way we live would be subject to considerable scrutiny and debate. But most of us don't see ourselves as influential participants who can offer informed opinions about the uses of technology. On the contrary, new technologies tend to be mystifying. They resist our attempts to get a grip on what they do and how they work.

As long as we think we do not have enough expertise to engage in substantive discussions about technology, we are effectively prevented from having an impact on the directions it may take. But there *are* opportunities for discussion and intervention in the process of technological growth and change, and it is important to take advantage of them. We believe that the lack of broad participation in conversations about technology seriously impoverishes the ways technologies are brought into our everyday lives. Our aim is to show how more people can be more fully engaged in important discussions and decisions about technology.

This book is a personal response to the prospect of increasing technological change. Our perspective comes from our experience as researchers in Silicon Valley and as users and consumers of technology. We, Bonnie Nardi and Vicki O'Day, have been trained in (respectively) anthropology and computer science. We have each crossed boundaries into the other's discipline during our years of working in industrial research labs, including those at Hewlett-Packard, Apple Computer, and Xerox. Both of us have designed and implemented computer software, and both of us have conducted empirical studies of how people use technology.

Our empirical studies are ethnographic studies, which means that we go out into the "field" to study situations in which people are going about their business in their own ways, doing whatever they normally do. For us, the field has included offices, libraries, schools, and hospitals. We observe everyday practices and interview people in their own settings over a period of time, to learn more about the complicated and often surprising workings of a particular environment. We bring the insights we develop from ethnographic studies to help in the design of technological tools that will be a good fit for the people who use them.

We consider ourselves critical friends of technology. We want to see more examples of good, useful applications of technology out in the world, like those we have seen in some of our studies. But as we do our fieldwork, read the newspapers, and watch the developments around us, we are sometimes troubled by what we see. Technical developments in everything from telephone menus to cloning and genetic engineering have potentially disturbing effects.

We have noticed that people seem to distance themselves from a critical evaluation of the technologies in their lives, as if these technologies were inevitable forces of nature rather than things we design and choose. Perhaps some of this lack of critical attention is due to the sheer excitement at the novelty and promise of new technology, which makes it easy to move ahead quickly and without reflection. For example, NetDays focused on wiring public schools for Internet access were carried out with good intentions, but we have seen that some of our local schools have had a difficult time coping with the new technology once they have it.

We are troubled when people ignore the human intentionality and accountability behind the use of technological tools. For example, when one of us recently forgot to pay a credit card bill and saw her credit card temporarily disabled as a result, she called her bank to ask it to accelerate the process of turning the card back on. She assumed that a twenty-year history as a good customer would make a difference. The response from each of the three customer service people she talked to was the same: "You know, you're dealing with a computer here." Well, not exactly. We are also dealing with people who solve problems and make decisions, or we should be. Human expertise, judgment, and creativity can be supported, but not replaced, by computer-based tools.

Many people have misgivings about technology, but most of the time we do not express them. Our own specific concerns are unimportant in this discussion. What is important is that each of us develop and use our own critical sensibilities about the technologies that affect us.

This book is addressed to people who work with and around technology. This includes schoolteachers and school administrators, engineers, salespeople, professors, secretaries, journalists and others in publishing, medical professionals, librarians, people who work in finance and banking, and many more. We believe that our colleagues in technology design will also find this book useful.

For all of our readers, what we hope to accomplish is a shift in perception. To explain what we mean, we can use visual perception as an analogy. Psychologists have studied the way people see, and recent research suggests that there is no conscious perception of the visual world *without paying attention to it.* That is, you can't see what is in your field of view unless you are prepared to notice and process what your eye takes in. According to Arien Mack and Irvin Rock, who have carried out research in this area, we are subject to "inattentional blindness" when we are not ready to pay attention to something in our field of view.[1]

This language about visual perception and inattentional blindness reso-
nates with our experience as researchers who study technology in use.
Sometimes we have seen different things in settings we have studied than
other technologists or even some of the participants themselves have seen.
We believe that to some extent, this is because we were prepared to see
and pay attention to different things. We were looking for a multiplicity
of viewpoints in the settings we studied, the hidden side effects of technology, people's values and agendas as they deployed technology, the
resources they brought to bear in getting their work done, the actual
work practices that accomplished the goals of the work, and the social
interactions that affected work and technology use.

In other words, some of what goes on in any setting is invisible unless
you are open to seeing it. We have noticed two blind spots people seem
to have in considering work settings: informal practices that support
work activities and unobtrusive work styles that hide valuable, skilled
contributions.[2]

In any work setting, there are commonly accepted accounts of the
regular and sensible ways things get done. They come in the form of
written procedures, job descriptions, organizational charts, project
planning documents, training materials, and more. While these descriptions are useful in helping people coordinate and carry out their work,
they do not always reflect the whole picture of the way things get
done. They capture the work activities, roles, and relationships that
are most visible, but not the informal practices that may be just as
important.

For example, informal collaboration among co-workers is common
but little discussed. An individual might own the responsibility for a
particular task, but behind this formal responsibility there are many
informal consultations and small communications with co-workers that
help get the task done. In engineering work groups we studied, engineers
asked one another for help in using complicated computer-aided design
tools, although coaching was not in anyone's job description.

Sometimes work is invisible because workers are intentionally unobtrusive in their activities. In our research in corporate libraries, for
example, we have seen that reference librarians usually carry out their
highly technical and skilled search activities behind the scenes, so much

so that their own clients are largely unaware of what they do. Since clients may not understand what librarians are doing, they may think that automated services can replace librarians. If they looked at the actual work, they would understand that automated services cannot perform the same tasks.

When new technologies are adopted into a work setting, they usually affect the informal activities and unobtrusive activities as well as the formal ones. As we plan the introduction of new technologies or the modification of existing technologies, it is useful to shift our perception and become aware of aspects of work that are usually invisible.

Though we are all subject to inattentional blindness, we can try to be more aware of informal and unobtrusive activities. By preparing to see between and behind the formal, well-advertised roles and processes, we can enlarge our vision. And if we learn to see our own settings differently, we will also be able to see different possibilities for discussion and local action.

THE RHETORIC OF INEVITABILITY

To achieve a shift in perception and prepare for conversations for action, we must look beyond some of the common rhetoric about technology. As we read and listen to what designers and technology commentators have to say, we are struck by how often technological development is characterized as *inevitable*. We are concerned about the ascendance of a rhetoric of inevitability that limits our thinking about how we should shape the use of technology in our society.

Some commentators welcome the "inevitable" progress of technology—that is the view of the technophiles, who see only good things in future technological developments. Some decry the inexorable advance of technology—that is the view of dystopians, who wish we could turn our backs on the latest technologies because of their intrusive effects on our social experience.

There are more possibilities for action than these extremes suggest. But to see past this pervasive rhetoric, we first need to bring it clearly into view, so we can recognize it, sensitize ourselves to it, and then move forward to a more fruitful position.

To consider just one of many examples of the rhetoric of inevitability, in an article in *Beyond Calculation: The Next Fifty Years of Computing*, Gordon Bell and Jim Gray of Microsoft assert that "by 2047 . . . all information about physical objects, including humans, buildings, processes and organizations, will be online. This is both desirable and inevitable."[3] It is instructive to read those two sentences aloud.

Humans are *objects*. We are in the same category as buildings. In this formulation, any special considerations we might have simply because we are humans (such as rights to privacy) are obliterated. The authors declare that creating a world in which people are objects in a panoramic electronic database is "both desirable and inevitable."

The authors use their authority as virtuoso engineers to tell us what they believe to be inevitable and to suggest how we should feel about it. Bell and Gray's article is not an anomaly. It is one example of many books and articles in which experts describe how technology will be used in the future with a sense of certainty, both about the technology itself and our own acceptance of the benefits it will bring to us.[4]

Bell and Gray state, "Only the human brain can process and store information more easily, cheaply and rapidly [than a computer]." The human brain is formulated here as cheap information storage. By reducing people's intellects to simple computation and storage capabilities, our goals and intentions and opinions are rendered invisible and uninteresting. We are concerned about the way the corporate mind is reaching into the future to define us as physical objects about whom any data can be stored online. Through the rhetoric of inevitability we are being declared nonplayers in the technical future. We are bargain basement commodities.

Another example of the rhetoric of inevitability can be found in the discussions of cloning people, which have featured inevitability as a constant refrain. Immediately after the story about the successful cloning of sheep in Scotland appeared in the newspapers in February 1997, a U.S. government spokesperson said, "Should we stop scientific development in these areas because the capacity [to clone humans] might become available? I don't think that's reasonable, or even possible. I just think that's one of the costs that come along with scientific discovery, and we have to manage it as well as we can. But it's awfully hard to stop it."[5]

The author of these remarks was Harold Shapiro, the chair of the National Bioethics Advisory Commission appointed by President Clinton. Surely someone appointed as a representative of the people's interests to advise the government on the ethics of biotechnology should take a little more time before declaring cloning technology inevitable. Is it not appropriate to have a public conversation about this far-reaching, controversial technology? Here the rhetoric of inevitability protects a scientific establishment that wants to be free of considerations of how its activities might affect the rest of society.

Shapiro was joined by Eric Juengst, a bioethicist at Case Western Reserve University, in declaring that banning future research is like "trying to put a genie back in its bottle."[6] The rhetoric of inevitability reaches a nadir in Juengst's comment: "Do we want to outlaw it [cloning] entirely? That means of course only outlaws will do cloning."[6]

There must be a better argument to be made about the implications of cloning than that only outlaws will clone if we make it illegal. Let's throw away all our laws, in that case! This is a sad logic, especially from someone described as in the newspaper as "one of the nation's leading biomedical ethicists."

Fortunately, the cloning discussion has been more polyphonic than many other technology discussions. In a story about cloning in the *San Jose Mercury News,* our local newspaper, it was reported that in 1973 the scientific community declared a moratorium on research in which DNA from one species was moved to another species, because there was popular concern about mutant strains of bacteria escaping from laboratories and infecting the entire world. In 1974, scientists urged the federal government to regulate all such DNA technology. Strict guidelines followed. They have been relaxed as the scientific community has taken time to sort through the issues and as public understanding has grown, but the regulations are widely regarded as responsible and socially beneficial steps to have taken at that time.[7]

Margaret McLean, director of biotechnology and health care ethics at the Markkula Center for Applied Ethics at Santa Clara University, wrote of the cloning debate, "We ought to listen to our fears." She noted that Dolly the sheep seems to be growing old before her time, possibly due to the aged genetic material from which she was cloned. McLean discussed concerns with attempts to overcontrol a child's future by controlling its genes, by setting expectations that a cloned child might find emotionally unbearable. She argued that we should consider our misgivings and give voice to them. McLean takes on the issue of inevitability squarely, declaring, "I, for one, believe that the possible is not the inevitable."[8]

The developer of the cloning technique himself, Ian Wilmut, voiced opposition on ethical grounds to applying the technology to people. There are already laws in some European countries that ban the cloning of human beings.

We hope that our readers will develop active antennae for sensing the rhetoric of inevitability in all the media that surround us. The cloning discovery and the variable responses to it show that there is not a single story to be told about any technology. Those who declare a technical development "inevitable" are doing so by fiat.

CONVERSATIONAL EXTREMES: TECHNOPHILIA AND DYSTOPIA

Conversations about technology are often positioned at one of two extremes: uncritical acceptance or condemnation. Writers of both technophile and dystopic works often assume that technological change is inevitable—they just feel very differently about it.[9]

These two positions stake out the ends of the continuum, but they leave us with poor choices for action. We want to claim a middle ground from which we can carefully consider the impact of technologies without rejecting them wholesale.

Nicholas Negroponte's book *Being Digital* is a shining example of the work of a technophile. Negroponte, director of the MIT Media Lab in Cambridge, Massachusetts, populates a new and forthcoming Utopia with electronic butlers, robot secretaries, and fleets of digital assistants.[10] In Negroponte's world, computers see and hear, respond to your every murmur, show tact and discretion, and gauge their interactions according to your individual predilections, habits, and social behaviors. Negroponte's lively future scenarios in which digital servants uncomplainingly do our bidding are always positive, unproblematic, and without social costs. There are some important pieces missing from this vision, though it is certainly engagingly presented.

Technological tools and other artifacts carry social meaning. Social understanding, values, and practices become *integral aspects* of the tool itself. Perhaps it's easiest to see this clearly by looking at examples of older and more familiar developments, such as the telephone. The telephone is a technological device. It is a machine that sits on a desk or is carried around the house, and it has electronic insides that can be broken. But most of us probably don't think of a telephone as a machine; instead, we think of it as a way of communicating. There is an etiquette to placing a call, answering the phone, taking turns in conversation, and saying good-bye, which is so clear to us that we can teach it to our children. There are implicit rules about the privacy of telephone conversations; we learn not to eavesdrop on others and to ignore what we may accidentally overhear. These conventions and practices are not "designed in" and they do not spring up overnight. They were established by people who used telephones over time, as they discovered what telephones were good for, learned how it felt to use them, and committed social gaffes with them.

Negroponte's scenarios are missing a sense of each technology's evolving social meaning and deep integration into social life. Though these social meanings can't be engineered (as the histories of earlier technologies have shown), we must understand that social impacts are crucially important aspects of technological change. We should be paying attention to this bigger picture, as it emerges from its fuzzy-grained beginnings in high-tech labs to saturate our houses, schools, offices, libraries, and hospitals. It is not enough to speculate about the gadgets only in terms of the exciting functions they will perform.

When we turn to writings in the dystopic vein, we find that concerns with the social effects of technology *are* voiced. But the concerns are met with a big bucket of cold water—a call to walk away from new technologies rather than use them selectively and thoughtfully.

A recent best-seller in this arena was Clifford Stoll's *Silicon Snake Oil*.[11] Stoll is an astronomer and skilled computer programmer who is well known for his remarkable success in tracking down a group of West German hackers who broke into the Lawrence Berkeley Laboratory computers in 1986. In *Silicon Snake Oil,* Stoll shares his concerns about the hype surrounding the Internet for everyday use. He suggests that consumers are being sold a bottle of snake oil by those promoting the Internet and other advanced technologies. In the rush to populate news-groups, chat rooms, and online bookstores in a search for community, we may find ourselves trading away the most basic building blocks for community that we already have—our active participation in local neighborhoods, schools, and businesses.

This is not an unreasonable fear. Another technology, the automobile, transformed the landscape of cities, neighborhoods, and even houses in ways that profoundly affect the rhythms and social networks of daily life. In the suburban Silicon Valley neighborhood where both of us live, each ranch-style house is laid out with the garage in front, making it the most prominent feature of the house to neighbors or passersby. The downtown shops are a long walk away on busy roads that are not meant for pedestrian traffic. Most people routinely drive many miles to get to work. We can be reminded of what our driving culture costs us by walking for awhile in a town or neighborhood built before cars—though this is not an easy exercise for Californians and other Westerners. In these earlier neighborhoods, there are mixtures of houses, apartments, and small shops, all on a scale accessible to people walking by, not shielded from the casual visitor by vast parking areas.

While we share Stoll's belief that the introduction of new technologies into our lives deserves scrutiny, we do not believe that it is reasonable or desirable to turn our backs on technology. It is one thing to choose not to use automated tools for the pure pleasure of doing something by hand—to create beautiful calligraphy for a poem instead of choosing from twenty or thirty ready-made fonts, or to play Monopoly (an activity advocated by Stoll) instead of Myst (a computer game with beautiful graphics). But sometimes the computer is exactly the right tool for the job, and sometimes it is the *only* tool for the job.

The issue is not whether we will use technologies, but which we will choose and whether we will use them well. The challenge now is to introduce some critical sensibilities into our evaluation and use of technology, and beyond that, to make a real impact on the kinds of technology that will be available to us in the future.

Stoll and Negroponte seem to be diametric opposites. Stoll says faxing is fine; Negroponte offers a withering critique. Stoll asserts that people don't have time to read email; according to Negroponte, Nobel prize winners happily answer the email of schoolchildren. Stoll tells schools to buy books; Negroponte says computers make you read more and better. But both Negroponte and Stoll are in agreement on one crucial point: the way technology is designed and used is beyond the control of people who are not technology experts. Negroponte asserts that being digital is inevitable, "like a force of nature." What Mother Nature fails to provide will be taken care of by the engineers in the Media Lab. And Stoll describes the digital promises as snake oil—not home brew. Neither Stoll nor Negroponte offers scenarios in which citizens have a say in how we want to shape and use technology.

A DIFFERENT APPROACH

Our position in this public conversation about technology lies between the positions exemplified by Stoll and Negroponte in some ways, and completely outside their construction of the argument in others. We share Negroponte's enthusiasm for and fascination with cutting-edge technology development. We share Stoll's concerns about the social impact of technology. But to shun digital technology as Stoll advocates is to miss out on its benefits. Neither does it seem wise to sit back passively waiting for the endless stream of amazing gadgets that Negroponte hypothesizes. It is not necessary to jump on the digital bandwagon. It is dangerous, disempowering, and self-limiting to stick our heads in the sand and pretend it will all go away if we don't look. We believe that much more discussion and analysis of technology and all its attendant issues are needed.

Some of this discussion is fostered by political action books, such as Richard Sclove's *Democracy and Technology*.[12] Sclove argues for grassroots political action to try to influence official governmental policies on technology. He writes, "[I]t is possible to evolve societies in which people live in greater freedom, exert greater influence on their circumstances, and experience greater dignity, self-esteem, purpose, and well-being."

We are in passionate agreement with this statement. At the same time, we recognize that politics per se—national, regional, or local policy advocacy—is not for everyone. There are other ways to engage with technology, especially at the local level of home, school, workplace, hospital, public library, church, and community center. We all have personal relationships with some of these institutions. We can influence them without having to change broad governmental policy, though that might happen in some cases.

In our research studies, we have seen examples of responsible, informed, engaged interactions among people and advanced information technologies. We think of the settings where we have seen these interactions as flourishing *information ecologies*. Each of these ecologies is different from the others in important ways. Each has something unique to teach us, just as we learn different things about biology from a coral atoll, a high desert, a coniferous forest. We suggest that these examples be read as stories that model a holistic, ecological approach to technological change. Using the metaphor of an ecology, we will discuss how all of us can find points of leverage to influence the directions of technological change.

3

A Matter of Metaphor: Technology as Tool, Text, System, Ecology

We have been talking about the ways that we have conversations about technology. We have visited dystopian and technophile perspectives, but there are other themes in other writings that we would like to turn to now. What we would like to highlight about these writings is the ways they use metaphors to reveal certain facets of technology.

Our concepts about technology are often embodied in highly packed metaphors. Metaphors are a useful form of shorthand. We believe it is important to look closely at these metaphors, because the way we use language conditions our thinking. The metaphors we discuss both illuminate and obscure the relationships between people and technology.

We use the term "metaphor" perhaps a bit loosely, but our goal is to suggest a sharp image that provides a neat encapsulation of a whole set of assumptions and questions. For example, people often talk about a technology as a *tool*. Because the purpose of tools is to accomplish something useful, talking about technology as a tool prepares us to think about the particular tasks people can accomplish with technology. The tool metaphor suggests that a person is in control of a technology she uses, because tools are objects we control. We often talk about technology as a tool in this book, because this usage is so common in the design community.

But it is important to recognize that all metaphors channel and limit our thinking, as well as bring in useful associations from other contexts. That is the purpose of a metaphor, after all—to steer us to think about the topic this way rather than some other way. Because we want to challenge traditional assumptions about how technology should be developed and used, we must also recognize and challenge some of the familiar metaphors that are commonly used to simplify and explain technology development. We will discuss in this chapter the metaphors of technology as *tool, text,* and *system,* to prepare for the discussion of our favorite ecology metaphor in the next chapter. Each of these metaphors, but most particularly the idea of technology as a system, raises the issue of autonomous technological change. Who is really in control of technology development? Is anyone?

For decades, public awareness of technological change has been colored by images of technology that has gone completely out of control, proceeding forward under its own strange logic, beyond the direct influence of individuals. This is in part what *Metropolis* depicted— autonomous machines exploded, malfunctioned, and finally led to the destruction of the city they were intended to support.

Lang's final images of technology were dark. But even when new inventions are received optimistically, our public discourse assumes that technological change moves ahead under its own steam, with no one in particular at the wheel. The sheep cloning stories we discussed earlier featured statements from scientists and government representatives that denied society's right to choose whether and how to proceed with cloning experimentation, including the nonsensical assertion that if cloning were outlawed only outlaws would clone. We ought to be able to do better than appeal to bumper-sticker slogans when we talk about fascinating and important developments such as cloning. Why do we resort to this ridiculous level of discourse?

Perhaps we find it more comfortable to assume that nothing can be done to change things than to face the question of what we should do. It is easier to believe that influencing technological change is someone else's business, and that our role is primarily to be appreciative consumers.

Public conversations and ideas about technological change also may be oversimplified because the big picture is overwhelmingly complicated. This complexity can be dismaying even for those who approach technology in the role of critical friend.

A whole school of thinkers in social and political theory, including Lewis Mumford, Jacques Ellul, Neil Postman, Langdon Winner, and Ivan Illich, have tried to come to grips with the interrelationships among technology and history, technology and social institutions, and technology and politics.[1] They point out that nothing about tool use is fundamentally new to us as a species, but that our ability to absorb new tools—and the different ways of *doing* and *being* that emerge with technological change—are challenged by the avalanche of innovation we are experiencing. Since the publication in 1954 of Ellul's masterful *The Technological Society*, social critics have sounded alarms about the stress to the human mind and soul of having to adapt constantly to the new. These scholars point to the erosion of tradition and identity entailed by the constant necessity of moving on to the next tool, the next technology, the next fundamentally different way of doing things. We are adapting to technology rather than controlling its fruitful and pleasurable use.

Ellul, Winner, Postman, and others present analyses and interpretations that are exhaustive, often very convincing, and extremely sobering. Readers are left with a formidably complex and essentially pessimistic picture that suggests few leverage points for purposeful human intervention. We will depart from the analyses offered by these writers in some important ways in the next chapter, but their construction of arguments about the potentially dehumanizing effects of out-of-control technology deserves careful consideration. Other writers (mostly insiders in the technology design world) are more optimistic about the effects of technology, but few of them address the systemic issues, the tightly woven interdependencies between technologies and their contexts of use.

Metaphors matter. People who see technology as a tool see themselves controlling it. People who see technology as a system see themselves caught up inside it. We see technology as part of an ecology, surrounded by a dense network of relationships in local environments. Each of these metaphors is "right," in some sense; each captures some important characteristics of technology in society. Each suggests different possibilities for action and change.

TECHNOLOGY AS TOOL

One way of understanding technology is to see each technology as a *tool*, something made to fit the hands and minds of individual human beings. This is the most commonsense definition of technology, and indeed "tool" seems to be merely a synonym for technology. But as we explore all the metaphors—tool, text, system, ecology—it will be clear that tool is only one way of looking at technology. There are other meanings of technology. Technology goes far beyond the notion of device-used-by-an-individual-to-get-something-done, which is the way we think of it in everyday terms.

The idea of technology as tool is so obvious that it hardly seems appropriate to attach a particular author to it, but for convenience in our discussion we can refer to Donald Norman's popular books, *The Design of Everyday Things, Turn Signals Are the Facial Expressions of Automobiles,* and *Things That Make Us Smart.*[2]

In these entertaining and enlightening books, Norman takes readers on a tour of everyday objects such as doors, cars, typewriters, faucets, and telephones. He shows how designers can help people figure out what to do with these objects by making the appropriate actions visible and obvious and by exploiting the inherent constraints of materials and things. If they are cleverly designed, objects can provide people with accurate clues about their use. This is the goal of the designer—we should not have to read instructions to know whether a door opens by pushing or pulling or to know how to turn on a faucet. It should be apparent that one ought to pull here and push there.

Norman's books are entertaining because he illustrates over and over again how often designers fail to achieve this goal. People will, in fact, get stuck behind doors they can't open and they will turn on the wrong burner on the stove. Designers may claim that users are quirky and users may claim the reverse; we only know that such design mishaps are commonplace.

The psychologist J. J. Gibson called intrinsic features of technologies (or of anything) *affordances.*[3] Gibson noted that we are pretty good at not falling over cliffs because we can see what would happen if we stepped over the edge. There is a strong visual affordance to a cliff edge. Gibson's original concept of affordances did not include much of a social context—he was looking at perception. But now we understand that many tool affordances have an important social dimension. We can think of affordances as those properties of an object that neatly support the actions people intend to take with the object.

If we think about tool affordances, a pencil provides a good example. It affords writing because it fits well in the human hand, it leaves an erasable mark, and the eraser is attached. The Roman alphabet has turned out to have interesting affordances for the electronic age—it is much easier to create electronic text with the alphabet because it is so small and general-purpose, compared to picture-based systems with tens of thousands of ideograms. (Although ideograms have other affordances, such as suggesting the meanings and derivations of words graphically.) It is interesting to look around to see what kinds of affordances tools have and do not have.

Some of a tool's affordances emerge during use, unanticipated by designers. Refrigerators are designed to keep things cool, but they also have a well-known affordance of providing a magnetic surface for hanging up notes, children's artwork, cartoons, and other family information. An industry of refrigerator magnet accessories has sprung up around the "extra" affordance—which is no less real than the affordance of cooling food.

Norman's books draw attention to the affordances of designed objects, and he offers valuable precepts to designers of tools of all kinds. As a psychologist, Norman naturally bases his advice to designers on how to support human cognition effectively. In his analyses of design successes and failures, he takes into account how human memory, perception, and reasoning work. He focuses on what happens when individual human beings interact with singular material objects—with little or no reference to the social situations or even the surrounding physical context in which these encounters take place.

Using the tool metaphor to describe technology suggests several tactics to users. Before starting to work, it is important to choose the right tool for the job. There is a matching process in looking at the task at hand and deciding on the best tool for that task. With the right tool, the task can be straightforward to accomplish. With the wrong one, it can turn into an amateurish mess. Once a good tool has been identified, it is important to learn how to use it well. In technology design settings we sometimes talk about "power users," talented people who can use "power tools" with great skill.

The tool metaphor also offers pointed suggestions to technology designers. Part of the delight of doing technology design is working with the materials (software, computer displays, networks) and making them do interesting, unexpected, and clever things. Thinking about technology as a tool helps designers remember that there is someone on the other end—people who are using the tool. It is not enough to think about the *tool's* inherent elegance and capabilities; one also has to think about the handles it offers to its users.

The area Norman has chosen to mine is rich and yields useful material. But it does not tell the whole story. We know that technological tools are embedded in a larger context, and that this context is important to understanding how tools are designed and used. For example, there are no design criteria that could emerge from Norman's tool-centric analysis that would guide developers of Web browsers in deciding whether or how to support filtering of selected web sites when the browsers are used in elementary schools. This is not a matter of efficient cognition, but a matter of providing tools to meet differing social values in differing contexts. The same could be said for encryption and its relationship to government control, the rights of individuals to keep information about themselves private, facilities for blocking email "spamming," and any issue that requires social as well as technical discussion. Physical context may also be crucial: how much space is there for equipment, can a child carry (and not break) a device, will there be enough phone lines?

Norman doesn't claim that designing tools is easy—on the contrary, he provides very convincing evidence that it is hard to do good design. He points out that it often takes a half-dozen design iterations before a new tool can be considered truly usable. But his discussion does not address issues that extend beyond individual human capabilities and cognitive needs. Some design problems originate in a larger context—the social, organizational, or political setting in which a tool is used.

We consider this larger context to be a legitimate focus of attention when we evaluate how technology works in a given setting. Evaluation should not be limited to cognitive issues such as whether menu items are easy to find or recognize, though these fine-grained questions must also be addressed. We would like to move beyond the human-machine dyad, expanding our perspective to include the network of relationships, values, and motivations involved in technology use.

The tool metaphor is useful for questions and discussions about utility, usability, skill, and learning. We need to keep this metaphor in mind, but we also need to look outside its boundaries.

TECHNOLOGY AS TEXT

Another very different way to understand technology is to see it as *text* —as a form of communication, a carrier of meaning that may be reinterpreted as the technology passes through different social situations. This is a favorite analytical approach of recent postmodern critical theorists such as Bruno Latour and Michel Callon.[4] There are many writers who theorize about texts, but we will only touch on the ideas of some to give a sense of how they are relevant to our understanding of technology.

Critical theory has changed the way social scientists look at commonplace objects and experiences. This perspective challenges us to see that a written text—a book, newspaper article, or other familiar written form—is not something whose meaning is stable, reliable, and created solely by the author. Every reader has an active role in constructing the text, to make it meaningfully present in the reader's own world.

A reader's construction is developed in a cultural setting that may be quite different from the writer's. As a result, the reader's understanding and appropriation of ideas can be very different from the writer's. The writer cannot command the unknown reader to develop one interpretation or another. He can only make suggestions within the language and structures of the text. He will succeed to the extent that his suggestions are intelligible and appealing to his readers.

It seems natural to extend this understanding of writers and readers to designers and the designed-for. The French sociologist Bruno Latour has written engagingly about the messages both sent and received through the design of technological artifacts. First, he points out that we can delegate certain functions either to human beings or to mechanisms. So, for example, we can delegate the task of closing a door behind us either to a human porter or to an automatic door-closing mechanism. Once a function has been delegated to a mechanism, the mechanism then "prescribes" to (tells) its users what they ought to do for the system to work properly. Consider the careful choreography needed to use a revolving door. You must scurry through the door at just the right moment to avoid a brief but embarrassing moment with a stranger in a small space, or the ignominy of watching several perfectly empty partitions rush past as you hesitate.

Ivan Illich, in a less playful mood than Latour, makes the same point about prescription, which he regards as an intrusive aspect of modern life. He says, "In . . . designed goods, the shape, color and provoked associations speak to the user about the way the item must be handled."[5] For Illich, there is something coercive and impersonal about a design intruding into our consciousness its information about "the way the item must be handled." In cultures where tools do not change quickly (perhaps not in a lifetime), tool use is learned in interaction with others, not through manipulations staged by unseen designers.

Prescriptions are written into technologies when they are designed. Prescription in the Latourian sense does not mean that a technology says once and for all how it will be used—or that it will be used at all—but rather that it makes claims on our attention in a particular way. Technological artifacts have a certain authority and presence.

Latour notes that "there might be an enormous gap between the prescribed user and the user-in-the-flesh."[6] When the gap between the designer's concept of the user and the actual user is too great, we end up with tools that are a very poor fit. In current design practice, engineers have few opportunities to interact directly with users. For the most part, the technology itself is the only medium for communication. As it moves from one group of people to another, it carries its own messages and meanings along with it—its prescriptions.

Textual analysis suggests different tactics than the tool metaphor does. Now we are encouraged to *read* the technology to understand its messages and imperatives. By thinking about technological artifacts as a form of communication, we can have a richer sense of the roles they play, and we can more easily look beyond a tool as a physical object to a tool as a kind of stand-in for other people who are not physically present. It can be easier to raise problematic social issues in the context of this metaphor. For example, questions of privilege (who decides which conversations in a video conference are recorded and where they are broadcast?) are more likely to arise here than in a discussion of a tool's functionality.

Technologies carry implicit suggestions from their designers, implementors, and purchasers about how to do things. However, these are only suggestions. There must be other active people in the picture too, who can modify or override the messages inscribed in artifacts. This metaphor of technology as text leaves us with an image of a clamor of voices, all bidding for attention. But what makes one voice louder than another? How do people decide among competing ways of doing things? How do people learn and change the way they do things over time? The text metaphor does not distinguish very clearly between talk and action. That is part of its point—it is useful to remember that talk is a form of action and action is a form of talk. But the metaphor doesn't tell us how people's judgment, creativity, and values can or should come into play when they choose to act. The text metaphor is useful as a way of prompting discussions of intentionality and meaning, but other discussions require further conceptual support.

TECHNOLOGY AS SYSTEM

Of all that we have read, the work that treats technology as system provide the richest, most troubling, and most mind-altering perspectives. If we are looking for breadth of vision, we can find it here. The complex systemic perspective taken in these writings yields provocative analyses of the pervasive influence of technology on our lives.

We gather these ideas together under the general heading of technology as *system,* but "system" here is not used as a metaphor of some familiar object or activity in the same style as tools and texts. The term is too big and vague for that purpose. Jacques Ellul, Langdon Winner, and others are attempting to bring together ideas about phenomena of immense scope, and the difficulty of this shows in the variety of metaphors, concepts, and examples they assemble. In their writing, they are clearly grappling to understand what it is they have their hands on.

There are important differences in the work of these writers, so it is in a sense unfair to lump them together this way, but what they have in common is a broad scope of vision and a deep concern about large-scale social and technical systems. Nor are they the only social commentators to present these ideas—we are focusing on their work as examples of a particular vein of thinking about technology.

Jacques Ellul, the late French sociologist and the earliest of the cultural critics we discuss here, presents a sweeping vision of technology in his book *The Technological Society* (in French, *La Technique,* first published in 1954). Ellul's argument turns on his notion of what he calls *technique.* There is no simple English translation for this term, so we will use the

French word. *Technique* is a cultural mindset in which pure, unadulterated efficiency is the dominant human value. This does not sound apocalyptic, but Ellul details, in relentless, chilling detail, how efficiency drives out every other human value. Technology comes into play because machines are so efficient; they are the standard of excellence in the world of technique. Everything else is to be compared to them. Everything—even people—evolves in the direction of mechanical efficiency. Ellul's most troubling argument is that technique is *autonomous,* that it proceeds under its own momentum without significant control by people.

Ellul defines *technique* this way:

> The term *technique,* as I use it, does not mean machines, technology, or this or that procedure for attaining an end. In our technological society, *technique is the totality of methods rationally arrived at and having absolute efficiency* (for a given stage of development) in *every* field of human activity.[7] [italics in original]

Ellul claims that his definition of technique is not a theoretical concept. Instead, he says, he has arrived at it empirically by looking at examples of human activity in many different realms and noticing how completely intertwined our technologies, social institutions, politics, and economics have become. Other writers, such as Winner, point out that these factors have always been intertwined, but they agree that certain fundamental changes in society have taken place since the Industrial Revolution.

Reading Ellul is a heady experience, not to be undertaken by the faint of heart. Ellul does not pull back from even the most thunderous proclamations. A few examples will provide a flavor of his themes.

Technique, says Ellul, leads people to adapt themselves to technological and institutional arrangements, rather than the other way around:

> Technique integrates everything. It avoids shock and sensational events. Man is not adapted to a world of steel; technique adapts him to it. It changes the arrangement of this blind world so that man can be a part of it without colliding with its rough edges, without the anguish of being delivered up to the inhuman. Technique thus provides a model; it specifies attitudes that are valid once and for all. The anxiety aroused in man by the turbulence of the machine is soothed by the consoling hum of a unified society.[8]

Technique is inextricable from people in society; it is now part of who we are:

[W]hen technique enters into every area of life, including the human, it ceases to be external to man and becomes his very substance. It is no longer face to face with man but is integrated with him, and it progressively absorbs him. In this respect, technique is radically different from the machine.[9]

Technique moves forward on its own, with new layers of technique building on previous layers with great speed, ignoring tradition that came before:

Technique has become autonomous; it has fashioned an omnivorous world which obeys its laws and which has renounced all tradition. Technique no longer rests on tradition, but rather on previous technical procedures, and its evolution is too rapid, too upsetting, to integrate the older traditions.[10]

Technique has its own rules and criteria and sets itself apart from moral values:

[T]echnical autonomy is apparent in respect to morality and spiritual values. Technique tolerates no judgment from without and accepts no limitation. . . . Morality judges moral problems; as far as technical problems are concerned, it has nothing to say. Only technical criteria are relevant.[11]

There are many similarly urgent passages in the brilliant but exhausting pages of *The Technological Society*. The tone of these assertions may cause readers to bristle, since there is little room here for ambiguity or hesitation. Ellul is operating without a safety net, and at times his argument seems precarious. He proffers no academic hedges, no protective disclaimers. Though it is tempting to dismiss his claims as exaggerated, we find that they do set up a resonance with our everyday experience.

For example, to return to the sheep cloning stories again, when we hear that we can't decide *not* to do cloning once we've learned that it can be done, we are hearing about autonomous technique in the practice of science, with its own rules and criteria that are distinct from moral values. The Scottish scientists who successfully cloned sheep find the idea of cloning human beings to be morally unacceptable and the United Kingdom has passed a law to that effect. But in the United States, the idea of applying moral principles to the process of scientific discovery was itself a topic of debate. There was no consensus on whether it was even appropriate to talk about ethics in this scientific research context.

The problem of applying moral values was demonstrated in the "cloning ban" enacted in June 1997 (the sheep were first cloned in February 1997) by a federal advisory panel reporting to President Clinton. The panel declared that it was "morally unacceptable" to clone human beings *at this time* "because it is neither safe nor effective."[12] The "morality" consisted of putting off the real decision about the ethics of creating human clones by narrowing the argument to one of safety and effectiveness—issues that will most certainly be resolved through further technical development. The panel acknowledged that cloning raised moral, religious, and ethical issues, but stressed that it did not base its recommendations on those grounds. It based its decision on the lack of safety of the cloning technique. "We have hung our hat on the safety issue," declared the head of the panel. On moral and ethical questions he said, "We have not said we favor one side or another." Indeed they did not; the ban applied only to a single cloning technique. Another very viable method is still legal.

Langdon Winner adopts a more measured tone than Ellul in his book *Autonomous Technology*. Winner uses Ellul's work as a foil against which to examine the meanings and origins of the idea that technology is out of control. While Winner finds himself exasperated by Ellul's breathless, everything-including-the-kitchen-sink style, he also comments that reading Ellul can keep you up at night worrying about what to do. Winner responds to three assumptions about autonomous technology that he and others find questionable:

- Technology is neutral; it is only how we use technology that makes it beneficial or harmful.
- People can control technology.
- People understand how technology works.

We will spend some time looking at Winner's responses to each of these assumptions, as a way of examining what Winner and others really mean by autonomous technology.

Neutrality of Technology

Most technology designers we know in Silicon Valley workplaces do not consider the new technologies they build to be especially laden with social agendas and values. The idea that technologies have social implications that developers ought to consider is even a little distasteful; it almost goes against the grain of free enterprise. Each new tool is a bid for a share in a very competitive marketplace—toss your hat in the ring and may the best (or most powerfully backed) widget win. If it is good, people will buy it, and if people buy it, it is good.

Developers do understand very well that technical merit alone does not guarantee success. The gamesmanship of product strategies is also an important factor. But basically, new tools are expected to be selected by customers on the basis of compatibility, functionality, performance, and innovation, rather than any implicit social policies of either the desirable or undesirable flavor. Surely it is up to purchasers, not designers, to decide how to use technological tools appropriately.

Langdon Winner refers to physicist J. Robert Oppenheimer's remarks in hearings before the Personnel Security Board in 1954 to explore this issue. Winner quotes Oppenheimer's reply when the latter was asked to comment on whether his qualms about the hydrogen bomb had increased as it became more feasible to build the bomb:

> I think it is the opposite of true. Let us not say about use. But my feeling about development became quite different when the practicabilities became clear. When I saw how to do it, it was clear to me that one had to at least make the thing. Then the only problem was what would one do about them when one had them. The program in 1949 was a tortured thing that you could well argue did not make a great deal of technical sense. It was therefore possible to argue also that you did not want it even if you had it. The program in 1951 was technically so sweet that you could not argue about that.[13]

As it became more and more feasible to develop the bomb, it became more and more irresistible to build it. Oppenheimer's feeling was that development was a good and sweet thing to do. He draws a clean line between development and use. Using the bomb was somebody else's business, not his. The development of technology is one enterprise, and its use is a separate enterprise that takes place in another world altogether.

Winner brings in Ralph Waldo Emerson's *Works and Days* to provide an alternative view. Emerson wrote: "Many facts concur to show that we must look deeper for our salvation than to steam, photographs, balloons, or astronomy. These tools have some questionable properties.

They are reagents. Machinery is aggressive. The weaver becomes a web, the machinist a machine." Technologies make demands on people. As reagents, they catalyze changes. Though no particular social agenda is sketched out here, Emerson suggests that technologies are not just objects of simple utility to their owners, since they also carry burdens of responsibility. Emerson (again as quoted by Winner) nails this point down with a homely example: "A man builds a fine house; and now he has a master, and a task for life: he is to furnish, watch, show it and keep it in repair, the rest of his days."[14] Anyone who has ever owned a house and attended to its needs will find this a convincing analogy.

Indeed, technologies are not neutral—at the very least, they invoke in us certain kinds of responses. These responses are not always pleasurable or advertised features of the technology, but they belong to that technology nevertheless. They are intrinsic features, not results that arise incidentally from use.

Technology *conditions* our choices, says Winner. We have an accumulated inheritance of previous technique, and we use this as an evolutionary starting point for any new forms we might come up with. Innovation certainly occurs, but within a well-established framework of preexisting material. Winner writes:

> [W]hen a badly constructed supertanker breaks up on the shoals, spreading oil on the beaches, we must understand that the event has something to do with decades of technical and social change that created the circumstances for the calamity. But does it make sense to say, as the voluntarist argument suggests, that we "chose" the design of the ships, the form of Onassis's corporation, the social and political conditions under which the boats sail, or the eventual crack-up? When we think back on it, we do not remember having been consulted.
>
> Much of our ordinary contact with things technological, I would argue, is of exactly this kind. Each individual lives with procedures, rules, processes, institutions, and material devices that are not of his making but powerfully shape what he does.[15]

Winner further describes a process of *reverse adaptation,* which is "the adjustment of human ends to match the character of the available means."[16] If our technologies only provide certain ways of doing things, and these processes lead to only certain kinds of results, then we will adjust our goals to aim for precisely those results. To follow Emerson's house example, if owning a house is the accepted way to get shelter, then people adapt to the requirements of home ownership and aspire to it. (Though Emerson's good friend Thoreau did show that there were a few alternatives, if one looked around.)

Winner also suggests that this goes beyond an adaptation of goals to a more pervasive adaptation of standards:

> We have already seen arguments to the effect that persons adapt themselves to the order, discipline, and pace of the organizations in which they work. But even more significant is the state of affairs in which people come to accept the norms and standards of technical processes as central to their lives as a whole. A subtle but comprehensive alteration takes place in the form and substance of their thinking and motivation. Efficiency, speed, precise measurement, rationality, productivity, and technical improvement become ends in themselves applied obsessively to areas of life in which they would previously have been rejected as inappropriate. . . . Is the most product being obtained for the resources and effort expended? The question is no longer applied solely to such things as assembly-line production. It becomes equally applicable to matters of pleasure, leisure, learning, every instance of human communication, and every kind of activity, whatever its ostensive purpose.[17]

The first question to ask of Winner's assertions is whether they are believable. The second is whether it really matters to anyone if they are. To respond to the first question, we do find this idea of reverse adaptation easily recognizable. We have encountered it in most situations in which we have studied or participated in the uses of technology.

Schools are a particularly fertile territory for finding examples. The "factory model" of schooling of the nineteenth century has been discredited in modern educational thinking, and students are no longer thought of as empty containers into which uniform knowledge can be poured to create uniformly good workers and citizens. But the education system still relies heavily on precise measurement of learning, as we can see from the current emphasis on assessing progress using standardized test scores.

Test scores and grades are understood to measure the amount a student has learned, and they are also used to rank schools. When parents go shopping for a new house, they may consult the average test scores of students in the local schools to decide how good those schools are. Test scores can even be compared among different countries around the world to find out which country ranks highest in math and science. We may take this for granted now, but as Neil Postman observes, the idea that thinking or understanding can be assigned a number, and that a measure of 88 reflects a different "amount" of understanding than a measure of 90, is a fairly recent invention. Examined in the cool light of our poststudent days, who can really believe these numbers? Yet they are firmly entrenched in the Ellulian technique of the school system.

Why should we care if we are participating in reverse adaptation to technology? If we have moved values of productivity and efficiency from the factory to arenas such as education where they did not formerly have much influence, does it really matter if we do this with technology?

It does matter, because the fit between the values we profess and the actual circumstances of our lives has a profound influence on that unmeasurable but very real thing, quality of life. At the very least we should sensitize ourselves to reverse adaptation and ask whether that is really the relationship we want to have with technology.

Controlling Technology

Technology optimists such as the MIT Media Lab's Negroponte clearly assume that technology is under human control. In the future, we will be able to buy robots, turn them on with perfect anticipation of where they will go and what they will do, and turn them off whenever we choose, after our expectations have been met. Right? Of the three assumptions Winner discusses, this is the one that is most likely to cause us to shake our heads.

We've seen both happy and unhappy examples of technology that gallops off to a surprising destination. A recent legendary example is the development of the World Wide Web and the consequent popularization of the Internet to millions of households and thousands of commercial businesses in an extremely short period of time. The Web originated in a simple software tool that was meant to allow Swiss physicists to share their publications more easily with colleagues in other regions. Many people helped make the Web more powerful and easier to use, but surely no one who contributed to its early development anticipated its phenomenal success.

Winner suggests that the real issue about control is that of unintended consequences, or what he calls "technological drift." We cannot possibly expect to predict or steer all of the results of innovation. In nontechnological areas that are not so saturated with visions of progress, we probably understand this better and would not expect to stay in complete control. The consequences of choosing a job, spouse, or place to live are unknown; some will turn out to be good and some not so good. We acknowledge to ourselves that each choice opens up some options and closes others, and we don't really know where it will all lead. The rhetoric about technological change tends to ignore the possibility of either unknown or negative side effects. This rhetoric inhibits our ability to examine our circumstances with a reflective eye.

An example Winner uses is the change experienced by the Sami of Finland when they adopted the use of the snowmobile in the 1960s. Though Sami did choose to use snowmobiles for herding reindeer, they did not consciously choose at the same time for the population of the herds to fall off because the snowmobiles placed too great a strain on the animals. They did not choose for the number of families who could be sustained by herding to be reduced, or for some families to drop out of the local economy and move away. As the economic imbalances grew, there were related changes in social status, moving from egalitarian to hierarchical social relations. The snowmobile led to a cascade of unintended and, in this case, undesirable consequences.

We need to learn to ask questions about the consequences of technological change in our lives, so we can seed opportunities for evaluation and correction and do a better job of noticing the early signs of unintended results. We also need to be much more critical about claims that the effects of a proposed change are well understood and beneficial, as in Bell and Gray's assertion that technological development as they describe it is "both desirable and inevitable," or Negroponte's assertion that technology is a "force of nature."

There will always be unintended effects of new technologies. Some of these unintended effects will be fortuitous and some less so. It is both misleading and patronizing to suggest otherwise to people who will live with the consequences of change. We do not mean to imply that change should be stopped, but rather that it should be expected and examined.

Every technology change needs critical friends to watch what happens, think about it, and provide useful feedback.

Understanding Technology

We may be able to understand the story of the Sami in Finland and their snowmobiles long after the chain of events unfolded, but surely the relationships between the technological intervention and its effects were not so clear at the time.

It is hard to understand technological change because the people and institutions involved are often large-scale and distributed. The diverse elements of a technical apparatus are too far apart in time or space or too enmeshed in complex relationships to grasp easily. Pollution creeps up on us because its contributing factors (cars, buses, barbecues, manufacturing plants) are so distributed in our economic and social systems that we can't develop a coherent action plan.

Ellul, Postman, and Winner are pessimistic about our ability to understand technology because they see it as autonomous in an invisible way—like the snowmobiles and the oil tankers and pollution. Ellul discusses the accretion of thousands of small decisions—made for the sake of efficiency *à la technique*—that lead imperceptibly to the dominance of technique. In Neil Postman's words, technology "has its own agenda."

Ellul argues that the development and deployment of technology take place in an inexorable monistic trajectory because the technical comes to dominate human thinking. This dominance occurs for two reasons. First, people are awed by technology and its apparently magical ability to work its effects upon the world. Second, and more pragmatically, technology becomes dominant because it *does* provide us with things that we value and that only technology can provide—powerful medicines and medical devices, climate control in our homes and workplaces, machines to reduce household drudgery, access to humankind's accumulated experience through books, film, computers, and many other media. So compelling are these treasures that it seems churlish to argue for any kind of technological restraint.

The process of technical dominance continues, says Ellul, as technology begets technology. The problems created by technology itself (say, a polluted river) are to be solved by more technology. The more technology we have, the more we will have, in a boundless kind of feedforward.

The systemic view of technology leaves us with sense of the inexorability of technological change. If we look at technology as a system in which we are all deeply involved, we can see why understanding technology choices and consequences is so difficult. The technological system is the water we swim in, and it has become life-sustaining and almost invisible to us.

Metaphors matter because they suggest particular avenues for action and intervention. Tools offer certain kinds of participation to their users, as texts do for their readers. What about systems—how do we understand the possibilities for conversations for action when we see technology as a complex system?

This is where we find the limits of the analysis of the technological system. This view does not address with enough force the possibility of local and particular change. Social critics identify sweeping problems, and they are naturally drawn to sweeping solutions. But the problems Ellul, Winner, and others depict are so large-scale and complex that it is hard to imagine successful solutions of the appropriate magnitude. Their proposals tend to be broad and systemic, just like the issues they identify. This makes their proposals next to impossible to implement.

On the whole, the social critics are much less convincing in their practical advice than in their superb analyses, even to themselves. They seem to be aware of the comparative weakness of their proposals. For example, Winner says, "I suppose I could fudge the matter here and seem to be zeroing in on some useful proposals. Having gone this far, the reader can probably predict how it would look."[18] Winner begins each suggestion with resignation and hesitation, "I could suggest . . . "

Despite his doubts, Winner makes an excellent contribution to the what-do-we-do question with at least one of his proposals, suggesting "the direct participation" of those who must use a given technology. This suggestion anticipates the arrival from Scandinavia of a practice called *participatory design,* in which workers and engineers collaborate in teams to design technologies for specific work settings, such as a bank or a bakery. It has been highly successful in Scandinavia and is migrating to other parts of Europe and to the United States.[19]

In the United States, however, participatory design is still primarily an experimental approach of researchers in universities or industrial research labs. It is not practiced in product settings where the goal is to develop widely used, shrink-wrapped software, for example. Ironically, product developers fear that collaborating with users in a few particular settings would make their software less generally usable by all—perhaps it is better to work closely with *no* users, so everyone will be at an equal disadvantage.

Winner also makes a "small is beautiful" argument, proposing that "technologies be given a scale and structure of the sort that would be immediately intelligible to nonexperts."[20] This echoes the principles of good design laid out by Norman and others. It makes sense for some technologies, such as ATMs (automated teller machines). An ATM is a networked device that handles a variety of transactions between different banking systems very quickly, yet it presents a simple, clear, limited set of choices to users who literally walk off the street. (Although haven't you had the experience of having to help the person in front of you with the ATM? Or maybe being that person!)

If we consider other everyday technologies, Winner's argument is less convincing. Do we understand the scale and structure of the phone system? Do we expect to be able to service our own cars or furnaces, install our own tires, troubleshoot problems in our personal computers? Can we drive a car without instruction? Even a bicycle is not "immediately intelligible." Supports such as training wheels and helpful parents are part of how we engage with the technology of a bicycle. As long as we are talking about simple user interfaces for routine functions, we are sympathetic to Winner's argument. But many technologies that we habitually use are well beyond our ken in terms of scale and structure. And most crucially, we rely on experts for certain aspects of our relationship with these technologies, such as installation and repair. So this argument goes only part of the distance in terms of helping us to figure out how to do better design.

Winner recommends that we "dismantle" (only for a short time!) those systems that seem troublesome, so we can study them better. "Prominent structures of apparatus, technique and organization would be, temporarily at least, disconnected and made unworkable in order to provide the opportunity to learn what they are doing for or to mankind."[21] Winner wisely makes no specific suggestions as to which systems we could actually dismantle. The motivations behind this idea are understandable from a scientific point of view. It is hard to decipher the smooth workings of a system when everything is going right—much easier in the presence of breakdowns, when the links between one part of the system and another are visible and more readily traced.

But this scheme is rather obviously impractical. Many of us have experienced the temporary "dismantling" of electricity, telephone, transportation, running water, and other technical systems during disasters such as earthquakes. What we learn is how much we long to have the taps flowing, the phones working, and the lights on. It is hard to imagine educators, politicians, and parents agreeing to dismantle the entire school system, for example, to see how it really works and figure out how to make it better.

Small experiments along these lines may be possible, though. That is what charter schools are about. In these settings, educators are exempt from many of the requirements of the institutional system and can find out how things operate without the moorings of the larger system. We will see in the decades ahead whether the lessons learned in charter schools make their way back to the many thousands of traditional schools.

In 1988, more than thirty years after the publication of *The Technological Society*, Ellul made two "proposals for action."[22] Like others, he issued a disclaimer before his proposals. Ellul warned that "[the proposals] will appear to you as highly utopian and illusory." They are certainly that, and the first is downright funny, too. Ellul's idea is to send almost everyone back to the countryside to produce high-quality food without agribusiness and chemicals. Ellul observes with a perfectly straight face that this must happen so that we can experience a return to "a good cuisine with good ingredients." He sighs, "You will no doubt say that this is very French." Despite the unquestioned virtues of French cuisine, we believe that something more is called for to address the problems of the technological society. Ellul's second proposal is to turn our productive apparatus toward producing "on a grand scale the basic necessities of life," which would be "furnished free of charge to the peoples of the third world." This is a noble sentiment, but Ellul does not extend it to any convincing practice or approach.

An idea common to Ellul, Postman, and others such as the late philosopher Michel Foucault is that we *resist* the technological society. We also encountered this stance earlier in Stoll's work. Over a period of many years, resistance has been developed as a major philosophical reaction to *la technique*.

Foucault has written extensively on resistance. He celebrates individual resistance to established power structures—feminists resisting demeaning medical procedures, anyone who resists ethnic, religious, economic domination or exploitation. As Mark Philip notes in a useful summary, Foucault champions "the sheer recalcitrance of individuals . . . and [Foucault's] works are intended as a stimulant to this recalcitrance."[23]

What Foucault and others mean by resistance is neither one-dimensional nor foolish, but still we find the idea of resistance both rhetorically and practically problematic.[24] Rhetorically, resistance *sounds* one-dimensional—with Luddite overtones, unintended though they may be. And "resistance" has a comical association with that well-known line from *Star Trek:* "Resistance is futile." This line has become a venerable cliché, settling into our collective consciousness in a way that makes it difficult to talk about "resisting" without eliciting smirks.

But more important than the fact that resistance is part of a pop culture formula is the fact that it is not clear according to what ethic one is to resist. Or if any ethic at all is to be involved. The idea of resisting seems to flow from a hidden but agreed-upon assumption that right-thinking people will resist in correct ways once they understand *la technique*. Maybe French philosophers know what to do, but what about the rest of us? Why resist? According to what benchmark? And if resistance means rejection of technology (as the term cannot help but imply), is it then impossible to enjoy the fruits and powers of technology in a responsible, reflective way? It is not uncommon to talk about throwing away one's television, but what about throwing away the phone and antibiotics? This is not likely to be a popular or practical idea. To resist, or rather, to merely resist, *is* futile.

As we look at the insightful analyses of social critics who understand technology as a system, we can hardly complain that this metaphor lacks scope, as we did with the tool metaphor. However, we can argue that the view of technology as system washes out the distinctions among different *local settings.*

In the next chapter we turn to our own ecological metaphor. We find the ecology metaphor powerful because it includes these local differences, while still capturing the strong interrelationships among the social, economic, and political contexts in which technology is invented and used. When autonomous technology is observed at the systemic level, its effects can seem overwhelming. But in individual local settings, we see a more varied texture of experience than we see from a distant vantage point. From the local perspective, we see paths toward creating reflective and purposeful uses of technology.

4

Information Ecologies

We define an information ecology to be a system of people, practices, values, and technologies in a particular local environment. In information ecologies, the spotlight is not on technology, but on human activities that are served by technology.

A library is an information ecology. It is a place with books, magazines, tapes, films, and librarians who can help you find and use them. A library may have computers, as well as story time for two-year-olds and after-school study halls for teens. In a library, access to information for all clients of the library is a core value. This value shapes the policies around which the library is organized, including those relating to technology. A library is a place where people and technology come together in congenial relations, guided by the values of the library.

A hospital intensive care unit is an information ecology. It has an impressive collection of people and technologies, all focused on the activity of treating critically ill patients. Human experts (nurses, physicians, therapists, ethicists) and machines (monitors, probes, and the many other devices in the ICU) all have roles to play in ensuring smooth, round-the-clock care. Though this is a setting with an obvious reliance on advanced technologies, it is clear that human expertise, judgment, empathy, cooperation, and values are central in making the system work.

A self-service copy shop is another kind of information ecology. In our local branch of Kinko's, for example, there is a floor full of copy machines, paper stock of different colors and patterns, paper cutters, scissors and glue, computers that can be rented by the minute, and laser printers and scanners. There is also a computer expert who sits on a stool near the row of computers to answer questions. There are workers behind the counter who can help with copying. Customers ask one another where to find supplies and how to get started on an unfamiliar machine. It is a busy and hospitable place.

In each of these settings, humans help other humans use technology. Simple things are done with simple tools. The library, hospital, and copy shop have typically sought out advanced technologies, but these technologies are carefully integrated into existing habits and practices, according to the values of the information ecology.

We introduce the concept of the information ecology in order to focus attention on relationships involving tools and people and their practices. We want to travel beyond the dominant image of the tool metaphor, an image of a single person and his or her interactions with technology. And we want to capture a notion of locality that is missing from the system view.

An ecology is complex, but it does not have the overwhelming breadth of the large-scale systems and dynamics Ellul and others describe. An ecology responds to local environmental changes and local interventions. An ecology is a place that is scaled to individuals. We can all name the ecologies we belong to and participate in. In an ecology, we are not cogs in sweeping sociological processes. Instead, we are individuals with real relationships to other individuals. The scale of an ecology allows us to find individual points of leverage, ways into the system, and avenues of intervention.

CHARACTERIZING INFORMATION ECOLOGIES

The notion of an ecology as we use it is metaphorical, intended to evoke an image of biological ecologies with their complex dynamics and diverse species and opportunistic niches for growth. Our purpose in using the ecology metaphor is to foster thought and discussion, to stimulate conversations for action.

We believe that the ecology metaphor provides a distinctive, powerful set of organizing properties around which to have conversations. The ecological metaphor suggests several key properties of many environments in which technology is used. An information ecology is a complex *system* of parts and relationships. It exhibits *diversity* and experiences continual evolution. Different parts of an ecology *coevolve*, changing together according to the relationships in the system. Several *keystone species* necessary to the survival of the ecology are present. Information ecologies have a sense of *locality*.

System

Like a biological ecology, an information ecology is marked by strong interrelationships and dependencies among its different parts. The parts of an information ecology may be as different from each other as the sand, sunlight, saltwater, and starfish of a marine ecology, but they are as closely bound together. In an intensive care unit, for example, the jobs of nurses and doctors can be seen to fit together in complementary ways, and the nature of their work is both extended by and dependent on the technologies they use in patient care.

Change in an ecology is systemic. When one element is changed, effects can be felt throughout the whole system. Local changes can disappear without a trace if they are incompatible with the rest of the system. For example, when schools set new goals for what students must learn in math, they also have to develop new ways of evaluating what the students have learned. Otherwise, teachers will find themselves under pressure to teach the material that was covered on the old tests, and the innovation will fail.

Diversity

In a biological ecology, different species take advantage of different ecological niches, which provide natural opportunities to grow and succeed. The complexity of biological ecologies ensures that there are niches for many different kinds of roles and functions. Just as it would be surprising to find only one grass or wildflower species in a biological community, we should not look for only one or two roles for people and tools in an information ecology.

In an information ecology, there are different kinds of people and different kinds of tools. In a healthy information ecology, they work together in a complementary way. In a library information ecology, for example, we find that librarians fill niches such as handling rare books, telling stories to children, answering reference questions, and publishing World Wide Web materials. All of these different roles of librarians help make the library work well for its community, providing different resources for varied audiences and their needs. The set of technical resources in a library is also diverse. There are computers that provide electronic catalogs and Internet access, paper and pencils for writing down call numbers, and labels on the shelves so you know which section of books you are looking at.

Diversity is necessary for the health of the ecology itself, to permit the system to survive continual and perhaps chaotic change. Monoculture—a fake, brittle ecology—gives sensational results for a short time, then completely fails. Information ecologies should be teeming with different kinds of people and ideas and technologies. It is captivating to wander through a rain forest and stultifying to be stuck in a hundred acres of soybeans. A diverse information ecology is a lively, human, intensely social place, even if it incorporates very advanced technologies. It has many different resources and materials and allows for individual proclivities and interests.

Coevolution

A natural environment offers many toeholds for life of various forms. With tenacity and vigor, species migrate and change to fill the available niches. These adaptations lead in turn to further change, as the entire system adjusts to new constraints and possibilities. A healthy ecology is not static, even when it is in equilibrium.

Similar dynamics are at work in evolving information ecologies. The pace of new technology development ensures that school, work, and home settings will continue to be offered newer, faster, and *different* tools and services—not just once, but repeatedly. Information ecologies evolve as new ideas, tools, activities, and forms of expertise arise in them. This means that people must be prepared to participate in the ongoing development of their information ecologies. For example, as schools across the country are wired by enthusiastic volunteers on NetDays, school teachers and administrators should expect to make decisions about how to use the new classroom Internet access not just once, but again and again. The Internet is rapidly changing, and the information ecologies in which the Internet plays a role must participate in those changes.

Information ecologies are filled with people who learn and adapt and create. Even when tools remain fixed for a time, the craft of using tools with expertise and creativity continues to evolve. The social and technical aspects of an environment *coevolve*. People's activities and tools adjust and are adjusted in relation to each other, always attempting and never quite achieving a perfect fit. This is part of the dynamic balance achieved in healthy ecologies—a balance found in motion, not stillness.

Evolution implies a past, as well as a future. An information ecology as a persistent structure over time acquires its own history. It displays the stable participation of an interconnecting group of people and their tools and practices. An experience with an ATM machine, for example, is not an information ecology. It is a useful but isolated service that is too simple to be an ecology. By contrast, a bank office is an information ecology with diverse services and activities, where there are interconnections among people and their tools. When we are in a bank, we can sense that the activities, materials, and tools of the trade have a continuing history of development and change.

Keystone Species

An ecology is marked by the presence of certain keystone species whose presence is crucial to the survival of the ecology itself. In the Indiana sand dunes, marram grasses send out root systems up to twenty feet in length to stabilize their sandy environs. Without these grasses and their roots, the dune sands would disperse and shift erratically in the face of strong winds blowing in off Lake Michigan.

When we add new technologies to our own information ecologies, we sometimes try to work in the absence of essential keystone species. Often such species are skilled people whose presence is necessary to support the effective use of technology.

Some high-technology businesses are recognizing the need for people who can serve as translators, facilitators, and teachers. For example, Farallon Computing, a California network products company, was

recently featured in a news article because of its innovative hiring practices.[1] This company has developed a strategy of hiring technical support
workers with little or no previous computer experience—a former
cocktail waitress, social worker, and hotel room service manager, for
example—because they outperform highly technical people in helping
other people with problems. As quoted in the news article, Farallon's
technical support manager said, "You can teach people to use a computer
but it's real hard to teach patience. I look for natural born teachers because that's what they're doing all day." Farallon has clearly recognized the
value of a certain keystone species—the natural teacher—in its work force.

Mediators—people who build bridges across institutional boundaries
and translate across disciplines—are a keystone species in information
ecologies. Ironically, their contributions are often unofficial, unrecog-
nized, and seemingly peripheral to the most obvious productive functions
of the workplace. Although the success of new tools may rely on the
facilitation of mediators who can shape the tools to fit local circumstances, technology is too often designed and introduced without regard
to the roles these people play.

Locality

In *A Midsummer Night's Dream,* Shakespeare wrote this description of
the creative work poets do:

> And as imagination bodies forth
> The forms of things unknown, the poet's pen
> Turns them to shapes, and gives to airy nothing
> A local habitation and a name.

We believe that a key to becoming an active participant in technological
change lies in joining ranks with the poets, whose creativity is grounded
in local settings. The notion of "a local habitation and a name" captures
for us the essence of an information ecology. The *name* of a technology
identifies what it means to the people who use it. In a sense, it positions
the technology more directly under the control of its users.[2] We do not
just refer to what the technology is called, but to how people understand
the place it fills. A computer in a library is most likely a card catalog or
an Internet access machine. A computer in an office is often a personal

information appliance. A computer in a small business might be a budget and payroll machine. In each of these settings the computer can have precisely the same hardware configuration, but what it *is* for each user population is different. This is not just a matter of different software packages installed on each machine. The identity of the technology is different in each of these local settings because the perceived role, availability, utility, and other properties of the machines are different. The local participants in each setting—librarians, office workers, small-business owners—construct the identities of their technologies through the rhythms and patterns of their use.

The *habitation* of a technology is its location within a network of relationships. To whom does it belong? To what and to whom is it connected? Through what relations? The habitation of a technology is its set of family ties in the local information ecology. An office computer is used by some person or group of people, maintained perhaps by others, and networked to other computers. It has a place.

We cannot overemphasize a key point here: only the participants of an information ecology can establish the identity and place of the technologies that are found there. Indeed, this is a responsibility, not just an opportunity. Designers of tools are responsible for providing useful and clear functionality, but they do not complete the job. As users of tools, we are responsible for integrating them into settings of use in such a way that they make sense for us.

Locality is a particularly important attribute of information ecologies. We all have special knowledge about our own local ecologies that is inaccessible to anyone outside them. Along with knowledge, we have influence. While it may be impossible even to think about trying to make an impact on national policy (unless you are a player in information ecologies at a national level—say, a member of the United States Senate or a federal judge), it is entirely possible to step up and say how you want to use technology in your own home, in your children's classrooms, at your workplace, in your doctor's office, or at your public library. These sites of local participation offer both opportunities and responsibilities for shaping the way technology works in our lives.

Only people who are immersed in a particular information ecology can provide a local habitation and a name to new technologies. Healthy information ecologies are sustained by the active, intelligent participation of the people involved in them.

WHY ECOLOGIES?

The word "ecology" is more evocative for us than "community," despite some similarities. Ecology suggests diversity in a way that community does not. Communities can be quite homogenous, or defined along a single dimension (the gay community, a community of scholars, a religious community). The parts of an information ecology are as different from one another as oak trees and scrub jays in a California woodland ecology.

Ecology implies continual evolution. The idea of community does not put the same emphasis on change. We often think (perhaps naively) of communities as timeless or slow to change (a prototypical Irish village or a Tibetan monastery).

There is an urgency in the notion of ecology, because we all are aware of the possibility of ecological failure due to environmental destruction. While communities do indeed break down, and there is anxiety about this breakdown, ecological breakdown is disastrous and irreversible in a way that community breakdown is not. We feel a sense of urgency about the need to take control of our information ecologies, to inject our own values and needs into them so that we are not overwhelmed by some of our technological tools.

Penetrating the process of technological development involves defining our own local information ecologies—creating a local habitation and a name for the technologies we use. Our leverage point lies in acting within the spheres where we have knowledge and authority. It may be that we will have the effect of shaping practice in our own settings with an extra measure of reflection and intention, or it may be that our efforts will be noticed and emulated by others. We are not asking people to think globally and act locally (recycling soda cans will not prevent Chernobyl). We are suggesting that people act locally in a committed, reflective way that acknowledges *technique* as Ellul documents it, but having recognized it, chooses to respond with initiative that is grounded in local understanding and values.

We cannot say how far this will take us. But we can imagine that if we used technology responsibly in our own homes, schools, offices, hospitals, libraries, and communities, a major change would be under way. The worst thing we can do is to ask too little of the future—and ask too little of ourselves in determining the future.

We believe that Ellul and other cultural critics are deeply pessimistic because their analyses are of whole systems—macrolevel processes that indeed seem impenetrable. As sociologists and political scientists, they are trained to look at the biggest picture possible. This is a wonderful gift and a very useful thing to do. But it is not the only way to see. Using anthropological methods and perspectives, we have looked "on the ground" at small social groups to find out what they are doing. Looking "in the small" can provide inspiration and practical ideas for how to change our own ecologies for the better. We see local participation as a viable point of intervention in a larger system that does, from many vantage points, seem to have its own agenda, as Postman says.

Rather than *resistance* we prefer to speak of *engagement* and *participation*—specifically, engagement and participation in our own information ecologies. Rather than individual heroics in the gynecologist's office or on the factory floor (à la Foucault), we advocate collective participation in socially shared and valued activities. While there are certainly times when resistance is appropriate, it is not enough, and we advocate taking the longer view involving the work of collective, ongoing *construction* of enduring information ecologies. The technology might be high or low; it should fit the needs of the ecology as determined by the members of the ecology.

Of course the problem of the pervasive, distributed technologies such as automobiles will continue to bedevil us, because they depend on a broad infrastructure and resist situation-specific adaptations. We must say more here about what we mean by participating in *local* information ecologies. "Local" is a relative term, relative to the specific individuals in an information ecology and their own spheres of *influence* and *commitment*. A head of state has a wider sphere of influence than a schoolchild. But each can speak up in his or her own ecology. Spheres of influence change over time, contracting and expanding. With pervasive communication technology, it is no longer appropriate to speak of a physical geography as providing a defining boundary (though it might). Local is now defined by influence in an ecology—which comes from participation and engagement—and commitment to a set of shared motivations and values.

Healthy information ecologies take time to grow, just as rain forests and coral reefs do. An information ecology begins with our own efforts to influence the shape and direction of the technologies we use and the settings in which we use them. We urge people to get involved in the evolution of their information ecologies—jump into the primordial soup, stir it around, and make as many waves as possible.

5
Values and Technology

In early 1997, pictures of Disneyland visitors in various stages of undress appeared on a number of Internet sites. A video camera mounted inside the Splash Mountain ride takes souvenir pictures of riders as they plunge down the five-story watercourse. The pictures are offered for sale to riders as they exit the ride. Some Disneyland veterans know when the photos are taken and pose for an "R-rated memento from the G-rated Magic Kingdom," as our local newspaper put it.

Disney owns the photos. Unbeknownst to Disney, a park employee (or employees) posted the pictures to the Internet. They have circulated at a number of sites. This incident raises the question: Is anyone's privacy being violated from the dissemination of the photos on the Internet? The park visitors willingly posed, they knew they would not own the images but would only be given a print of the photo, and they were aware that park employees would see the photos.

While Disney could sue the perpetrators, they have not chosen to do so. The rationale is that no money is being made from posting the photos. And indeed, why not post the photos more publicly? Where exactly are the boundaries of privacy to be drawn? Does Disney have an obligation to provide privacy for its visitors? Or should visitors assume that if their picture is taken, it is fully public? With the ubiquity of video cameras and the possibility of easy publication of images to a worldwide audience, privacy issues are more complex than ever.

The Disney example is an amusing instance of much more serious privacy problems raised by technology. For example, should the Internal Revenue Service and the Department of Motor Vehicles share data with other government agencies and private companies, as they have discussed doing? Should we be videotaped at work? When are audio and videotapes admissible in court? How about genetic privacy? Do employers and insurers have the right to our genetic data?

Many values are challenged and renegotiated with new technologies. Values such as autonomy, privacy, freedom of speech, copyright, accountability, and universal access come into play in the technological realm. Indeed, our deepest ideas about what it means to be human are confronted when we consider technologies such as cloning or those used for third-trimester abortion. We do not want to argue for any particular stance on any particular issue here, but rather point out that values are intimately involved in the everyday choices we make in using technology. Technology engages everything from the frivolity of Splash Mountain to our most cherished beliefs about life itself.[1]

We view values not as immutable, clearly defined objects, but as negotiated processes. Free speech, for example, is a value most Americans agree upon, but it remains an area of active debate and interpretation. In the area of technology, because we are experiencing such rapid change, there is no set of established and well-understood values that come with the territory. The processes of negotiating values are crucial. Many school districts, for example, are engaged in the process of trying to figure out how to give students access to information over the Internet (since a core value in school is encouraging students to seek information themselves), yet avoid unsavory material or time-wasting treks across cyberspace. Hospitals must find a way to allocate scarce medical technologies. Businesses have to negotiate issues of employee privacy with email and computer monitoring.

Langdon Winner drew our attention to the idea that technology is not neutral. It invites certain responses by strongly suggesting what we can and cannot do. Technology does not determine our decisions, but it isn't exactly an innocent bystander.

By the same token, people are not neutral either. We bring our values to bear in designing and using technology. The key constituents of an information ecology—people, practices, values, and technology—exist in relations of interdependence. As we course down Splash Mountain, we decide how to respond to the possibilities of the technology. Likewise, when we sit down in front of the computer to think about what to post on the Internet, values come into play. There is no irresistible force that leads to the taking of sexy pictures (in fact, who would have thought. . . ?), nor is there any compelling need to broadcast pictures to the world. Values shape our actions.

Every business, school, or other organization has some motivating agenda, a reason for being. Although values may or may not be openly discussed, certain values are likely to be associated with that agenda: integrity in selling insurance policies, broad outreach in community college education, compassionate care in medical practice. These values can be applied when new technologies are considered. Healthy information ecologies rely on the integration of technology decisions and the values associated with the local setting, whether it is an intensive care unit, a classroom, or a retail store.

We raise the whole issue of values with some trepidation. Values discussions are often taboo or regarded with disdain, for a couple of good reasons. First, people think you want to dump yours on them. This is certainly a legitimate concern. However, we are arguing for local participation, local engagement with technology, which grows out of local conditions unique to specific settings. The values of a neonatal intensive care unit, for example, emerge from the activities peculiar to the concerns of taking care of very sick babies. Local values may relate to other more widely held values, but they are unique in reflecting the particular setting in which they are found and enacted in everyday activities.

The second problem with values discussions is that it has been legitimately asked whether we have any left. Neil Postman, Jacques Ellul, and scholars in the communitarian movement believe that we have been all but stripped of our values by the rapid pace of technological change and other concomitant social changes. We must ask ourselves if we actually have any values worth injecting into our local information ecologies.

In his book *Technopoly*, Postman argues forcefully that we have lost track of fundamental values.[2] One of Postman's most passionately held beliefs is that we have few values left because our culture is receding under the wave of an "information glut." Postman makes a powerful and disturbing argument that the information glut is rapidly diminishing the social institutions, art, and cultural values that have sustained us for centuries. According to Postman, information technologies are largely to blame.

Postman's argument goes like this: With so much information inundating us, all of it becomes meaningless. We are barraged with words, images, and sounds at such a high rate, with so little coherence and meaning, that we cannot maintain a cultural center of gravity—we cannot sustain a coherent set of values. We can no longer follow a normal process of absorbing and responding to information because it is here today and gone tomorrow.

Postman says there is too much information coming too fast. We are flooded with information from geographically distant places to which we have no real connection. The mass media feed us dramatic decontextualized "highlights" (often with visceral visuals) from this war-torn zone, that scene of a massacre, this group of starving children, that political scandal. What do we really know about the places this "news" is pulled from? We are not meant to make a connection to the human tragedies or follies that we witness—how many of us know what became of the people of Biafra, Eritrea, Armenia, and Iraq? We are not meant to achieve a contextualized view of life in, say, Iraq, to develop a relationship with a locale that lasts longer than the current crisis. We are only meant to tune in to the ratings-rich moments of collapse or annihilation, or to indulge an "inquiring minds want to know" voyeurism.

Postman notes that the technologies that first permitted the barrage of information were telegraphy and photography. The telegram and the photograph made possible the decontextualization of words and images. Nuggets of information contained in photograph and telegram-sized packages from all corners of the globe could now be rapidly and cheaply disseminated. The cost of information plummeted. Penny presses, newspapers, magazines, and posters proliferated. The symbolic environment was strewn with ceaseless, irrelevant information. As Postman says, "[I]nformation appear[ed] indiscriminately, directed at no one in particular, in enormous volume and at high speeds, and disconnected from theory, meaning, or purpose."[3]

Postman credits Ellul with developing the basic ideas that inform his views. If we go back to Ellul and see what he has to say about values, we find him at his bleakest:

> [H]uman beings are already modified by the technical phenomenon. When infants are born, the environment in which they find themselves is tech-

nique, which is a "given." Their whole education is oriented toward adaptation to the conditions of technique. . . . and their instruction is destined to prepare them for entrance into some technical employment. Human beings are psychologically modified by consumption, by technical work, by news, by television, by leisure activities (currently, the proliferation of computer games) . . . all of which are techniques. In other words, it must not be forgotten that it is this very humanity which has been pre-adapted and modified by technique that is supposed to master and reorient technique. . . . We are no longer . . . dealing with an ethics of choice with regard to possible futures. Choices and orientations in technique are made according to technical criteria and not in virtue of some deliberate human decision which has been made as a choice between several non-predetermined possible solutions. In the same way, any reference to values . . . is meaningless since the values defined by the traditional societies no longer have anything in common with the use of technique.[4]

All right, somebody had to say that, and Ellul said it well. However, this is a deeply pessimistic, claustrophobic view of human life. It is good to grapple with this view, but we must also come up for air and look around. When we do that, the situation just doesn't seem as desperate to us. Despite Postman's compelling examples and Ellul's no-holds-barred laments, we think their arguments exhibit too much sweep and too little nuance. We do not believe that information technology has crushed and eliminated all other values.

We have only to look around to see evidence of a diversity of values in public life. For example, many countries have strong, enforced laws against child labor, formerly a common practice nearly everywhere (and still common in some places). Some people are working hard to put pollution controls and worker safety measures into place. In the United States, there has been a revolution in rights for disabled people. By arguing for and applying values (beyond those obvious to our economic system, such as efficiency and profit), we have a society in which children do not have to work, where workers are guaranteed a level of safety, where we are at least attempting to clean up industrial pollution, and where the disabled have some measure of assistance. Concern for the many constituencies represented by these examples springs from a diversity of values and their active application in public life. Within economic life, there is scope for thoughtful values, such as when consumers make purchasing decisions based on a company's environmental friendliness or when companies advertise their ethics to add clarity and appeal to their identity. In several of our empirical chapters, we will highlight the way values come into play in different information ecologies.

Let's consider Ellul in a better mood. He wrote movingly:

> We must not think of the problem in terms of a choice between being determined and being free. We must look at it dialectically, and say that man is indeed determined, but that it is open to him to overcome necessity, and that this *act* is freedom. Freedom is not static but dynamic; not a vested interest, but a prize continually to be won. The moment man stops and resigns himself, he becomes subject to determinism. He is most enslaved when he thinks he is comfortably settled in freedom.[5]

This feels a lot better. Ellul puts the responsibility for action squarely on us, but at least in this formulation there is a possibility to take free action. If we feel discomfort at aspects of the technological society, if we are uneasy, questioning, probing, then we are freer than if we harbor the illusion that choice is an uncomplicated business in which we can become "comfortably settled." Freedom is a personal and cultural accomplishment that vanishes when we resign ourselves, as Ellul says. Certainly it vanishes when we accept the rhetoric of inevitability.

In *Faust,* Goethe said much the same:

> To this opinion I am given wholly
> And this is wisdom's final say:
> Freedom and life belong to that man solely
> Who must reconquer them each day.

We are responsible for reconquering freedom each day in our information ecologies, as they come into existence and evolve over time with our active, thoughtful participation.

The application of human values in information ecologies brings to bear a different dynamic than that of biological ecologies. We make deliberate, conscious choices about how we want our values to influence practices and technologies in information ecologies. There is a complex dance between two nonneutral forces at work here: technology with its texts and affordances, and people with their values and choices. The choreography of the dance is up to the human side of the equation, but only if we choose to "overcome necessity" by engaging our values and commitments as we shape our information ecologies.

6

How to Evolve Information Ecologies

We have been urging readers to get more involved in their local information ecologies. We believe that there are some practical ways to do this effectively. They can be summarized as follows:

- Work from core values. In *Metropolis* terms, work with heart, as well as head and hands.
- Pay attention. Notice what meanings are assigned to technologies as they are used, or intended to be used, in your ecology. Reflect aloud about what you notice.
- Ask strategic, open-ended questions about use. Perform thought experiments by asking "what-if" questions along the way.

There is a Zen saying that, roughly paraphrased, embodies this interesting idea: when you consider an object, it is what you *see* that makes the object beautiful and what you *don't see* that makes it useful. A bowl has a visible shape, color, and texture to admire, but it also sculpts out an empty space that gives the bowl its utility—this is the space where you can actually put something into the bowl. A guitar or other musical instrument has a wooden form with particular shape and markings, and a hollow interior that gives the sound resonance and amplification. Here is a poem that expresses the idea:

> Thirty spokes will converge
> In the hub of the wheel;
> But the use of the cart
> Will depend on the part
> Of the hub that is void.

> With a wall all around
> A clay bowl is molded;
> But the use of the bowl

Will depend on the part
Of the bowl that is void.

Cut out windows and doors
In the house as you build;
But the use of the house
Will depend on the space
In the walls that is void.

So advantage is had
From whatever is there;
But usefulness arises
From whatever is not.

We work with being
But non-being is what we use.[1]

We think this Zen notion applies in information ecologies too. The part we often focus our attention on is the technology: computers, networking, applications, handheld information gadgets, instruments, monitors, widgets ad infinitum. We look at the shape, color, texture, and functions of the technologies, and we think creatively about how to make them more usable, appealing, and effective. But it is in *the spaces between these things*—where people move from place to place, talk, carry pieces of paper, type, play messages, pick up the telephone, send faxes, have meetings, and go to lunch—that critical and often invisible things happen. As we look at information ecologies, whether they are examples in this book or examples from our everyday lives, we need to be mindful of those spaces.

Here's an example of someone looking clearly at the useful spaces:

> My dream is that for every medical device or technology that is developed in this country, at the same time we'd have a whole team of creative people focused on, "OK . . . what are the kinds of services we can give people [using this technology] to get them back on their feet?" This type of work isn't going to win any Nobel Prizes. . . . The Nobel Prize might be for humanity.
>
> —Catherine Hoffman, senior policy analyst, the Kaiser Family Foundation[2]

Hoffman is envisioning the spaces between the medical devices that preoccupy us—the spaces that could hold crucial services that, if offered, would "get people back on their feet." In this pithy statement Hoffman makes several key points: (1) the services are just as important as the devices; (2) the services require the talents of creative people; (3) we do not value this kind of creativity the way we value more purely technical creativity; and (4) service provision requires a team approach in which work and creativity are distributed across people, not concentrated in the heroics of a single genius.[3] Hoffman has seen, with Zen clarity, into the interstices between devices and geniuses, a space where we have the opportunity to provide greater utility "for humanity," as she says, through the provision of services.

Of course what constitutes the "spaces" in any given information ecology varies. Our point is to look for the spaces, to consider not just the obvious (the beauty of the physical that attracts the eye) but the potentials of the spaces—much less obvious but ultimately the most useful.

Healthy information ecologies are characterized by technology use in a social matrix consisting of services, norms, and conventions. These establish appropriate usage, core values, support, and a growth path for users that helps them become more competent with technology over time if they so choose. These social practices are an important element of diversity in an information ecology, providing not just the actual technologies themselves, but ways to use them.

WORK FROM CORE VALUES

Whether we are talking about tools or practices, core values are the center of gravity of a healthy information ecology. Lang stressed the importance of values in *Metropolis,* and he suffered withering criticism because of it. The story line and archetypal characters of *Metropolis* are certainly tidier and less realistic than we are used to in today's films, but Lang's powerful message about the importance of the human heart has always been correct. There is no basis other than human caring and love for deploying technology humanely. Lang's own words, quoted from an interview he gave toward the end of his life, underscore the simple truth of his film:

> I didn't like *Metropolis* after I finished it because I didn't think in those days a social question could be solved with something as simple as the line: "The mediator between the brain and hand must be the heart." Yet, today

when you speak with young people about what they miss in the computer-guided establishment, the answer is always: "The heart!" So, probably the scenarist Mrs. Thea von Harbou [who collaborated on the screen play] had foresight and therefore was right and I was wrong.[4]

The development of healthy information ecologies must rely on values (in addition to considerations of productivity and effectiveness), to avoid internal contradictions that can lead to failure. In some of the cases we focus on in later chapters, we will see, for example, that issues of access equity are extremely important to schools that are introducing technology to students for the first time. It matters whether poor students can use technology as readily as those who are more well-to-do, and whether girls use computer-based tools as readily as boys.

Information ecologies are systems of parts that fit together well—and the idea of "fit" must be understood in terms of social values and policies, as well as tools and activities. If the practices that evolve in a sociotechnical system are efficient and productive but fail to uphold the ideals or ethics of the people involved, the system will be subject to considerable stress.

In every setting in which we can see technology being introduced (workplaces, schools, libraries, hospitals, government agencies, homes), people have beliefs and assumptions about their rights and responsibilities. But our society has such a strong consumer orientation that people are not always eager to apply their principles as they make decisions about design or use. We have heard arguments that the marketplace will decide what is appropriate—technology designers can simply populate the marketplace with as many new ideas and tools as possible, and let Darwinian selection take its course. Instead, we believe that we should explicitly refer to values as we evolve our information ecologies.

PAY ATTENTION

One of the easiest traps to fall into as we consider our work practices and technologies, and the ways they might fit together congenially, is to take either a technology or work practice for granted. There are two foolish mistakes we can make: we can assume that the way something is now is the way it has always been and must be, or we can assume that the way something is now has no particular motivation or rationale behind it. Paying attention means deliberately evaluating whether a practice or technology has merit, and if so, what this merit consists of—what does it mean within a particular information ecology? It is tremendously valuable to wonder about why things are the way they are. It is even more valuable to *reflect aloud* about what has been noticed so that others can take part in the discussion.

Here is a concrete example. At a school in the San Francisco Bay area, the teaching staff, parents, and school administrators debated the appropriate uses of email. For some of the experienced email users in the group, the discussion at times seemed extremely tedious. Who would not find email useful? What possible rationale could there be *not* to adopt email to help with all kinds of information sharing in the community? The agendas and minutes of board meetings could be distributed through this medium at lightning speed. The weekly newsletter could be sent through email, so there would be no more lost copies as it was passed from teachers to forgetful children, on its way home to parents.

As community members (those who were on the email list, of course!) discussed possibilities and concerns, some interesting observations emerged. One of the most important principles of this school is sharing responsibility and information with children. Wouldn't something be lost if they were out of the loop of the weekly newsletter? Few of the children have email accounts of their own; only their parents have access to email. If the weekly newsletter were just read by parents online, it might no longer be posted on refrigerators for children to see and consult for themselves. Another important principle of this school is the building of relationships through personal interactions. Board meetings can include long, frank discussions of troubling issues. How much easier *should* it be to get instant access to detailed meeting minutes in an impersonal delivery format? Is there not some benefit in going to the administrators' office to get the minutes, since useful conversations might start with the people in the office who had attended the meeting? As the online debate continued, these observations of current practice were eye-opening for people who had originally seen the email question in fairly simplistic terms. Ultimately, the community decided not to distribute the newsletter or board meeting minutes on the email list, but to leave it as an informal discussion channel.

This particular school has a technologically literate and extremely cautious user community. In this case, it helped that the community had a high comfort level with the technology in question. They already knew and loved email in other settings, and they did not have to prove to anyone that they were on the cutting edge. Their familiarity gave them the courage to slow down and pay attention.

ASK STRATEGIC QUESTIONS

The people in this school community did not just ask *how* to use email. They also asked *when* email was appropriate, *why* they should choose to use it, and *why* they shouldn't. It is common to leap ahead to "how" questions when we think about technology. There are many difficult practical issues to resolve: how to pay for it, how to make space for it, how to accomplish certain tasks using a new technology, and how to train people in its use. It is less common—but crucially important—to ask a full range of "why" questions as well: why we are led to use technology for this task, why this particular technology seems best, why it fits well with our current practices or why it doesn't, why it will be a good idea to change our current practices. "Why" questions explore motivations, objectives, and values, while "how" questions focus on logistics and tactics. Unless the "why" questions are answered, the greatest skill in addressing "how" questions can still result in a misguided technology implementation. Asking both "know-why" and "know-how" questions is the crux of strategic questioning.

At the elementary school that one of our children attended, the new principal recently shut down the computer lab. The lab had been wired by enthusiastic NetDay volunteers, equipment had been donated by local businesses and parents, and parent volunteers had helped gather software and peripherals to round out the lab's offerings. The principal, however, asked what the purpose of the lab was. She pointed out that there was no one to manage the lab, and there was no tie-in to the school curriculum. In short, there was a room full of equipment but no information ecology (though of course she did not use our term). The principal faced sharp criticism from some parents, but she was courageous, in our eyes, in asking the know-why questions.

There are several kinds of useful questions for participants in information ecologies, including the what-if questions (thought experiments) parents and staff used in the email debate and open-ended questions (without a fixed set of possible answers) like "Why are we doing this?" These questions cannot be answered with a simple yes or no; they do not encode the questioner's viewpoint as thoroughly as closed-ended questions do. For example, it is more powerful to ask "Who can contribute to technology training for each classroom?" than to ask "Will district trainers or master teachers do the technology training for each classroom?" The first question is more open-ended, and it might lead to unorthodox possibilities such as involving parents or students in training.

Fran Peavey, a San Francisco social activist, has used the term "strategic questioning" to describe the process of asking questions in order to create strategies for action.[5] The question format she describes is easily adaptable for participants of information ecologies—indeed, we expect that many people will recognize in these questions some they have already discussed in their own ecologies.

There are several question families. Some of these questions lead to important conversations about know-why, helping us get beyond our fetishized attention to know-how. Why indeed do we want to engage with a particular technology? These questions also draw on our familiarity and expertise with our own *local* settings. There are no standard, off-the-shelf answers. The answers must arise from the particular local circumstances of each ecology.

Some strategic questions may lead to taboo topics, raising hidden agendas to the surface or revealing tensions in current power structures. Though some issues that arise during technology discussions can be uncomfortable to discuss openly, it is far better to discuss them while there is time to influence the direction of the technological change.

For each question category we have provided examples of what people in a local school ecology might ask, in order to illustrate the questions with concrete examples, but the questions apply to any information ecology. We chose the school example because of our own familiarity with school settings, through our experience as researchers and parents.

Local knowledge is needed to decide which questions are important to ask, as well as to explore their answers.

Questions that Describe the Issue

- **Analysis questions,** which are concerned with motivations, opinions, and relations between things.
Example: What is the real goal—what do we hope to accomplish with technology in school?
Example: Who should participate in forming and carrying out a technology plan?
Example: Is the technology intended to be pervasive throughout the school or associated with different grade levels, curriculum areas, or particular teachers?
- **Observation questions,** which elicit what can be seen and heard.
Example: What do students and teachers say they want to do with technology?
Example: How are computers or other technology resources currently used in each classroom? At home, by students or teachers?
Example: How much space is available for computers in classrooms, the library, and other places in the school?
- **Focus questions,** which identify the situation and some of the key facts.
Example: What human resources do we have available to help with technology—knowledgeable staff, parents, students, or local businesses?
Example: What technology resources do we already have in place now?
- **Feeling questions,** which relate to emotions and health.
Example: What successful and unsuccessful experiences do the teaching staff consider they have already had with computers in the classroom?

Questions that Dig Deeper

- **Visioning questions,** which ask people to identify ideals, dreams, and values.
Example: If we could do anything, what would be great?

Example: Where would we like to be in five to ten years?

Example: What values related to learning do we bring to this planning process?

Example: What do we value in the professional practice of teaching and how might the introduction of technology reinforce or revise the professional practice?

• **Change questions,** which look at how to get from the current situation to the desired situation.

Example: How do we want to change the way students learn? Which students are most in need of extra help in our school?

Example: How will we fund technology purchases?

Example: How will we fund maintenance, troubleshooting, and curriculum consultation?

Example: Are we willing to depart from the technology plan to take advantage of unexpected opportunities, or should we be firm about sticking to the plan?

• **Questions that consider alternatives.**

Example: Should parents and staff work together on technology planning and implementation, or should we have separate groups to permit more focused planning and brainstorming in each group?

• **Questions that consider consequences.**

Example: If we decide to invest in technology, what will we *not* have time, energy, space, or money to do?

Example: Do the people who have expertise to contribute to technology development have the freedom to take up new responsibilities? Will they be able to drop other responsibilities in exchange?

Example: How will the introduction of technology affect the way the teaching staff is recognized and rewarded for its performance?

Example: When the technology plan has been implemented, will there be an alignment between the teacher's authority to manage the classroom and the teacher's power to affect the workings of technology in the classroom? What will be the relationship and accountability between the teachers and technology developers and support staff?

• **Questions that consider obstacles.**

Example: What kind of training must be provided for people who have interest but lack background?

Example: What about teachers who have no interest in technology—can they skip using it in their classrooms?

• **Questions that take personal inventory** of interests and contributions.

Example: What different kinds of expertise are needed for the technology plan?

Example: What parents, students, teachers, and community members can we draw on to help us reach our technology goals?

Example: What partnering or mentoring relationships are already in place or should be set up to help with the plan?

• **Questions that elicit specifics** of what to do and how and when to do it.

Example: How will we plan to share experiences—both successes and failures—regularly, so we can learn from each other as the plan is implemented? What forums do we need for this purpose?

Example: When and how will we evaluate the process? How will we know if we are succeeding with the plan?

Example: How will we communicate our findings to the broader community?

The most important idea to remember about strategic questions is that local knowledge is required both to *formulate* and *respond to* questions, and local knowledge is distributed throughout an information ecology. No single person can know enough to ask all the right questions. A diverse set of perspectives is needed to develop a healthy information ecology. This means that everyone should be encouraged to ask questions, not just those with highly visible technical knowledge or management responsibility.

COMPLEXITY AND CHANGE

Developing a sound information ecology is a complex business. Each person, tool, and practice is mutually dependent on many others, and there are a variety of perspectives and concerns in any single ecology. When faced with this complexity, there is a danger of becoming "analytic paralytic," as one creative problem-solver put it. As we pay attention, ask questions, and come to understand tools and practices in more detail, all of the possible directions of change can become confusing and overwhelming. How are we to decide which actions to take?

There are two responses to the problem of analysis paralysis. The first is to recall that the process of discussion—just talk—has the power to change things. Thought experiments, in which people imagine a particular change and its consequences, that are carried out publicly can help people clarify some of the puzzling possibilities. Second, even small-scale experiments can be extremely valuable. A teacher, for example, might try out an idea in one classroom, using what she knows about herself and her students, rather than start by developing a technology plan for the whole school. Local experimentation can provide guidance to a larger-scale planning and implementation process.

There is a powerful synergy between changing tools and practices. As people become more involved in their own information ecologies, they will be able to articulate more clearly and precisely what works and what doesn't, what they value, and what they need and want. They will be more cognizant of the possibilities in technology and more creative in pushing it forward to meet their own needs.

The locality and access we all have to our own information ecologies are key to resisting the seeming inevitability of technological change that appears to be beyond our control. While a local information ecology might seem too small a leverage point, consider these words of Margaret Mead: "Never doubt that a small group of thoughtful, committed citizens can change the world; indeed, it's the only thing that ever has." Margaret Mead left us with many gifts; one of the most valuable was her words on commitment and change. As we confront the rhetoric of inevitability and the steamroller of technological change, it is good to remember Mead's optimism and her own unique, committed contributions.

In the next section of the book, we look in detail at specific information ecologies. In each chapter, we draw out the ecological characteristics of the local settings we investigated. Looking at schools, hospitals, offices, and libraries, we see examples that show diversity, coevolution, keystone species, and the application of values.

II
Case Studies

7

Librarians: A Keystone Species

In *The Diversity of Life*, the biologist Edward O. Wilson writes about why biodiversity is important and what happens when it is lost.[1] Biodiversity is threatened in many biological ecosystems, because of the impact of changing human activities. Wilson believes that we can slow this trend, but first we need to survey these ecosystems to find out which species are present and, most important, to understand their contributions. With this understanding, we can create sustainable biological wealth.

Diversity should be preserved, Wilson says, for at least two reasons. First, every species alive today is a natural survivor that may offer as yet unknown benefits for our own health and productivity. Second, and most important, an ecosystem as a whole is threatened by the loss of certain species. If we do not respect diversity, we may unintentionally damage the entire ecosystem. Wilson writes:

> Field studies show that as biodiversity is reduced, so is the quality of services provided by the ecosystems. Records of stressed ecosystems also demonstrate that the descent can be unpredictably abrupt. As extinction spreads, some of the lost forms prove to be keystone species, whose disappearance brings down other species and triggers a ripple effect through the demographies of the survivors. The loss of a keystone species is like a drill accidentally striking a powerline. It causes lights to go out all over.[2]

The wedge-shaped stone at the pinnacle of an arch—the keystone—stabilizes the arch and holds it together. Like the keystone, certain species in an ecosystem are crucial to the shape and stability of the system. In *Natural History*, Yvonne Baskin described the painstaking fieldwork conducted by biologists to understand how keystone species provide vitality to ecosystems.³ For example, along the intertidal rocks on the coast of Washington state, starfish prey so successfully on mussels that they keep these aggressive creatures from monopolizing space on the rocks. This leaves room for limpets, barnacles, and other marine organisms. Keystone species may literally sculpt the environment so that a variety of organisms can be hosted. Elephants' foraging habits, for example, prevent shrubs and trees from dominating the African savanna, providing a home for hundreds of different species of birds and animals. Gophers, badgers, rabbits, and ants tunnel and churn the soil, aerating it and loosening it so that it provides a congenial habitat for a plethora of species. All of these species are keystone species, central to the robust functioning of the ecosystems of which they are a part.

Biologists point out that you can't identify a keystone species with a casual glance—species do not wear their indispensability on their sleeves. Biologists undertake extensive fieldwork to identify keystone species and show exactly what the species are contributing to their ecosystems.

This chapter is about the value of diversity in a particular kind of information ecology, a library. Libraries offer a surprising variety of services and resources. Many individual clients of libraries are unaware of this breadth, since their preferences and needs have led them to experience only part of what the library offers. We show a broader picture of the library than many individual clients see.

Because of the development of new information technology on the Internet and elsewhere, this is a time of change for libraries and library services. Sometimes evolution in ecologies is not slow and gradual, but abrupt. When a volcano erupts and drastically changes the landscape, some species never make it back and others return in new roles. The growth of the Internet is like a volcano in our midst, and we should expect radical changes as a result. Just as word processing programs on personal computers have largely replaced typewriters, online information access will replace some of our old "manual" approaches.

More information is becoming available online, and more people are becoming adept at using online search tools. It is legitimate to ask if the need for libraries and librarians is diminishing. As do-it-yourself information finding in cyberspace becomes more common, will librarians go the way of custom tailors—unusual and perhaps a little antiquated?

Off-the-shelf clothing doesn't fit anybody as perfectly as custom-made clothing would, but it is good enough (and cheap enough) for most of us. Perhaps the same could be said for off-the-shelf information services.

We believe that the diverse services available in the library are still important and useful, and we believe that the increase in online information presents more opportunities to leverage the skills of professional librarians than ever before. Through our own fieldwork in libraries, we have identified librarians as a keystone species. Librarians' contributions are vital to the success of the library, and their roles can be expanded very usefully into cyberspace. By observing how librarians work in their most common physical setting, the library, we can envision some of the ways they can apply their skills and knowledge to Internet-based information services. At the same time, their presence continues to be important in physical libraries.

Because many people are generally unaware of how libraries work, there is a real temptation to assume that the librarian's work is easily automated. Although librarians offer services that are unique and crucial to the robust functioning of a library ecology, some of these services are invisible and easily overlooked during planning (and budget cutting). Certainly there is a trend toward more and better tools for online information access. Digital library technology seeks to digitize as much information as possible. Accompanying this development is a need to enhance information-finding tools such as keyword-based search interfaces or software programs called "agents."

Keyword searching is probably a familiar idea to most people. The information seeker types in one or more words of interest. Then powerful "search engines" proceed through the collection of documents to see which ones contain those keywords, and the documents that match are presented to the user. (That's the basic idea, though keyword-based searches can be very sophisticated.) Agents are small, focused software programs that can do the same or similar searches repeatedly. They typically do one thing well, such as finding all the articles about your ten favorite mutual funds and performing that search each month for the next six months.

As new, powerful search tools become more widely available, they are helpful additions to an information finder's tool kit. Digital libraries offer exciting possibilities for information sharing and learning. But we do not believe that these technologies should drive out librarians. The classic ecological pattern of invasion and succession that can transform biological ecologies so radically and rapidly is all too possible in information ecologies. A much better approach is to encourage mutual adaptation, fostering new relationships between the technologies and the practices of librarians and people who are trying to find information.

Our goal in this chapter is to highlight the diversity of the library information ecology and describe some of the ways its different contributors work together. To understand the diversity of the library, we must look in some detail at how its services work, just as biologists do when they go into the field to examine the workings of a biological ecosystem. We cannot take a snapshot of an ecology and hope to see how the pieces fit together. We have to look at the ecology in action and over time.

Many of the contributions of librarians are invisible to library clients— not by accident, but by design. It is actually part of the professional practice of librarians to protect their clients from the messy details of their work. As a result, few people have much of an idea of what librarians do. We will go behind the scenes to describe some of the invisible work that takes place in the library.

The descriptions in this chapter are based on two separate studies we carried out at the Hewlett-Packard Library and the Apple Research Library.[4] These are both corporate libraries where company employees can seek information related to their work. The libraries have materials related to engineering, physics, medicine, chemistry, business, and other topics, and they have interlibrary loan privileges with a network of other libraries. They provide superb resources to their clients.[5]

Both of these studies were ethnographic studies—that is, we went to the places where reference librarians worked and we observed and interviewed them. We also interviewed library clients in their own work settings and looked at examples of searches they had asked librarians to help them with. We attended a short course on online searching for professional searchers. Altogether, we spent several months in and around these libraries, developing a sense of what went on in each place. All of our interviews were recorded and transcribed, and throughout the chapter we will intersperse excerpts of our recorded conversations, to bring to life the ideas of the librarians and library clients we met.[6]

Since this is the first chapter that dives into a particular information ecology, we raise an important general point. When we are trying to understand something about the way an information ecology works, it is vital to gather real examples from daily practice. Each example may look like an exception to the "rule," but that is part of the point. It is far too easy to make generalizations that sustain common fictions about the way things work, smoothing out differences and idiosyncracies. Real examples show diversity and interconnections that summaries often conceal. When people give general accounts of how things work to outsiders, they usually leave out what is locally unimportant or little valued. But sometimes these hidden details, seen from different perspectives, emerge as crucial to the workings of the ecology.

THE LIBRARY ECOLOGY

What belongs in a library ecology? In a reference library or in the reference section of a public library, some of the basics are information resources (books, magazines, videos), tools to help people use these resources (catalogs, computers, indexes), reference librarians, and people who come in to find things. The tools have changed over the years, as new technologies have become available. The Apple and Hewlett-Packard libraries were early adopters of new technologies such as online information services, CD-ROM information bases, and the World Wide Web.

In both of these libraries, self-service tools were available. Clients could browse online card catalogs, online collections of periodicals, or the World Wide Web just as they would have looked through stacks of books and magazines a few years ago. When people enjoyed browsing or when they were not sure how to describe what they wanted until they saw it, they often did their own looking without needing help from anyone else.

When someone is looking for a particular kind of information in a library rather than simply browsing, the task can be easy or difficult. It is easy when you know the title and author of a book—you can look in the card catalog to find the book's call number and then find it on the shelf fairly quickly. Or when you need to know something straightforward and factual, you can look it up in an almanac or encyclopedia. Librarians call these "ready reference" searches.

Sometimes finding information is harder, and you don't know where to start looking. An enormous amount of information is available in reference libraries, both on paper and in electronic form. It can be a formidable task to find the right places to look and then pinpoint what you want when you are looking in those places. Reference librarians are consultants who help people find information. They are especially helpful in working through the hard, poorly defined searches, where part of the goal is figuring out how to frame the question.

In our fieldwork, we saw that the consulting reference work of librarians was deliberately carried out in the background. Part of the service they provided was to give unobtrusive guidance, neither pushing too much information at one time nor leaving their clients to stew. Librarians moved back and forth between low-key conversations with their clients and expert, rapid information probes, in which they often managed to extract just what the client was looking for. In their conversations with clients, librarians tried to bridge gaps in language and expertise between an individual who had a particular interest in, say, learning about telecommunications standards, and the people who publish information relevant to that interest—such as researchers, engineers, legal experts, mainstream and specialized journalists, and government policy experts.

To expose what is often invisible, here is a brief description of what reference librarians did in the two corporate libraries we studied. They received requests for information by phone, fax, email, or in person. They interviewed clients to find out what those requests really meant. (Could the client really want everything ever written about plastics? Probably not.) They created search plans and then carried them out, sifting through both electronic and paper materials, or they taught their clients how to use the various databases and indexes to carry out searches themselves. Librarians paid attention to the intermediate results they were getting and they followed the trails that seemed most useful and appropriate for the client. Then, when the client had learned something and had follow-up questions, the librarians often helped with the next round of searching.

Library clients, librarians, and software tools work together. Each has something important to contribute. The client knows the background of the request—he knows why he's looking, what he already understands, and what he's going to do with the new information. The librarian knows what kinds of information sources are available, which tools can be used to access particular information sources, and what kinds of results other people have found useful in similar circumstances. The software tools provide the hooks that are needed to locate and pull out particular information of interest from large collections.

The more challenging information-finding tasks can only work out well when all three of these parts of the library ecology (librarians, clients, and tools) combine their capabilities harmoniously. To illustrate the close relationships between these different members of the library ecology and their importance to the health of the ecology, we will describe in more detail some of the roles librarians play.

INFORMATION THERAPY

One of the most valuable (and unheralded) services librarians provide is to help clients understand their own needs—a kind of information therapy. Interacting with a reference librarian can be very much like going to a good psychotherapist who, through skillful questioning, gets you to talk about what's really bothering you. At first you only know you feel bad (you need some information), but you can't articulate the issues in a precise way. After some discussion with the (information) therapist, the problems become clearer and more focused. Librarians are professionally trained to carry out these "reference interviews"—a modest name for an impressive deployment of tact, diplomacy, and persistence, as well as a skillful interviewing technique.

Broad, imprecise, and vague search requests are a common occurrence for reference librarians. Here are some of the actual requests received by librarians at Hewlett-Packard and Apple, taken from our studies:

"I want to know about plastics."

"I want to look at how other companies have set up Science Advisory Panels."

"I want everything on [some company]."

"I want to know about the presence of HP in Japan and in Europe."

"I need articles on Unix."

"Well, I have come across this organization and I have no idea what they do. Could you see if there is anything on them?"

"I'm trying to get a list of all books available for Windows, PCs, and Macs."

"I want all the articles written by a guy I met at a conference three years ago, but I've forgotten his name."

Try to imagine how you might deal with these queries yourself, if someone asked you for help. Even if you had ideas about how to proceed, some of these requests very likely don't capture in their current wording what the client really wants. (*All* the books on Windows, PCs, and Macs? A guy whose name you can't remember?) It is important to spend some time reformulating the information request before moving forward. This reformulation begins in the reference interview, a low-key interaction between the librarian and the library client.

The librarian uses the information in the client's original request as a starting point in the interview. The interview becomes a back-and-forth negotiation in which the librarian tries to learn what the client really wants. From the client's point of view, this feels like a brief, informal conversation—most clients probably are not aware that they are being "interviewed." From the librarian's point of view, the interview is focused on eliciting essential details about the client's needs, details the librarian knows will make a difference in carrying out the search.

The librarian might ask how recent the materials should be; whether summaries or full articles will be more useful; about how many items will feel like the right number; and whether the search should be comprehensive or just a good sampling. She will probably suggest a few vocabulary words and ask questions that stretch the boundaries of the search space, just to make sure that the client gets what he needs. She has technical expertise in searching the appropriate databases, indexes, and catalogs, so she knows which pieces of additional information can help in focusing the search.

The reference interview is the librarian's first opportunity to work with the client to shape the request into a plan that can actually be carried out. In the interview, the librarian tries to clarify and focus the search, gathers a sense of what amount and kind of information the client expects to find, and probes to see if any related information might be useful. In this conversation, the client and the librarian *construct together* a clearer picture of what the information request is about. The picture could not have been produced by either the client or the librarian alone. This is a creative, interactive process, to which each person brings special expertise and knowledge.

Here a librarian described some of her client interactions, with requests ranging from the specific to the vague.

> It varies all the time. Some people are very specific and say, "I need these articles," or "Here is the topic, here are fifteen keywords that are related to it, I need information from this year." Other people come and say, "Well, I have come across this organization and I have no idea what they do. Could you see if there is anything on them?" So, it can range and, you know, some people know exactly what they are looking for and others want anything that just happens to pop up or anything related to something that pops up. Usually it is up to me or whoever the person is who is going to be doing the searching to ask, "Okay, what does that mean? Is there anything else I can search on if I don't find anything under that?" That sort of thing.

To clarify a search, the librarian considers different interpretations of the search request. She uses her prior knowledge of the client's activities or direct questions to choose among the possible paths. Here is how one librarian responded to the request to search for information about the "presence of HP in Japan and in Europe":

> First of all, having heard the request, see what the person meant by Japan vs. Europe. Do they mean market share, the number of units? Do they mean plant size, relative presence? Do they need something that is more economic, which might be conversion ratios? And when you finally hit the one that you think they need, you maybe take a couple others.

In this example, the librarian raised alternative interpretations the client may not have considered. In the reference interview, librarians often make a significant contribution to defining the search they will eventually carry out on a client's behalf.

Focusing a search is often necessary, especially when clients' requests would result in more material than they could possibly want. In response to the client who wanted "all books available for Windows, PCs and Macs," the librarian told us about the careful negotiation she conducted to focus the search:

And I said to him, "Well, do you know how many this is going to be?" So I got him to narrow it to books that are currently in print so we would get rid of out of print books. And then books from the last two years. And I got him to narrow it to titles, only. . . . Actually, I had to go back and renegotiate with him to narrow it to the last year. And it came out to a list of 1,006 books.

One way to broaden or narrow a search is to select the keyword search terms, dates of material, and information sources very carefully. Does the client want historical background, because he or she is new to the topic, or only the latest developments? Does the client want articles for experts or novices, scientists or businesspeople? What about regional, national, or international perspectives? Is the client seeking articles that are centrally focused on the topic of interest, or articles that may have only peripheral references to the topic?

Librarians often push the boundaries of a search, just to see whether something else might be useful. For example, one librarian said that whenever she receives a request for all the technical papers by some author, she asks whether the client wants the author's patents as well. Most people do not think of searching patent databases to learn about new developments in a field, but there is good technical information in these databases that may never appear in journals or news articles.

Here a library client described how his librarian expands his searches:

She knows me fairly well and knows what I want because I have been working with her for now three years. But she tends to go overboard which is good because she always gives me more information than I wanted and does the extra little 20% that she tends to do so well, and she will usually call me back and say, "Jerry, okay, I have done the research, but here is some more information—do you want it or not?" Usually I'll take it, because she went through the effort of getting it anyway. She tends to ask about more detail if she is unclear about exactly what I am looking for, and if she has found something else out there that kind of keys into what I've been saying in a different way, she will also follow up and say, "Well, this is what I found and I found this too, do you want that?"

In many of the searches we analyzed, there were extended negotiations between the librarian and the client as the client's goals were sharpened and refined. The librarians helped clients *understand their own needs.*

It is very important to see that information requests may not be vaguely expressed just because the information seeker can't come up with the right words to describe them. Some requests are vague in their expression because that is the kind of information problem they are at the moment, for the client. People actually learn more about what they want by going ahead with the search, ill formed as it might be. Going ahead is the only way to make progress and understand better what the information needs really are.

This is a useful observation to keep in mind when we evaluate the features of information search tools. Clever vocabulary selection will not make open-ended information problems any less vague. The software tools should be able to do something reasonable with loosely characterized requests that result in an avalanche of information, as well as offering users ways to express very precise requests. In general, to understand how an information ecology works we need to turn from people to their tools and back again—to see what each is doing and how information is expressed and exchanged at different times.

We have talked about how librarians work to understand the client's goals. Sometimes librarians deliberately violate a client's goals, even after they understand them. This happens when two conditions are met: the librarian perceives that it is in the client's best interest to do something other than what the client has asked for, and it does not seem worth negotiating a new goal. For example, one librarian described how he did a search in which the search criteria as given by the client were too specific to generate results:

> There weren't enough articles to use these kinds of criteria. That happens fairly often where they [give] very specific criteria. [In this case there were] only five articles [that matched with looser criteria], so to hell with the criteria—just give 'em the five articles, and they can decide if they're worth anything or not. [And] you know, if the date range he gave was 1992–93, and I only found two articles, [I would go back further and let him know], "Well, this search covers 1991–93," just to indicate that maybe I violated his criteria a little bit, but hopefully in his interest.

The librarian understands that there may be a difference between what the client says literally and what the client actually wants. If clients' requests were always carried out exactly as presented, they would often yield poor results.

Sometimes clients try to meet the librarian more than halfway, specifying a search request in terms of a few very specific keywords. In the following exchange, the librarian described an overly specific request to the interviewer, who did not yet understand how librarians worked with such requests:

Librarian: (*reading the request out loud*) "I need a list of books with the following words as keywords."

Interviewer: Oh this sounds like a straightforward one. OK.

Librarian: Well actually those are the ones we hate.

Interviewer: Yeah? So what does the rest of it say?

Librarian: "I need a list of books with all three words in the title or as subject: connectors, contacts and reliability."

Interviewer: And, why—this looks like a classic one to me. How can you hate this?

Librarian: It's one of those that they are trying to get halfway into the process. This person is obviously looking for information on something related to connector reliability. OK. First of all, if I just did this search the way he had asked for it, chances are I would not come up with anything related to what he wanted. He wants all three words in the title, or as subject. If I just went in with those words in the subject, those words would not appear in any of the LC's [Library of Congress's] subject headings.

Interviewer: Because?

Librarian: Because that's just not the way LC works!

Interviewer: Right. OK, now is this a controlled vocabulary issue?

Librarian: That's a controlled vocabulary issue. They just don't do it that way. Yeah, they don't have that specificity in their controlled vocabulary; the language is a lot more archaic in the Library of Congress setup because it's these things are historical artifacts they don't want to change controlled vocabulary. So if something is set up in 1975, that is the terminology they are going to use until that becomes so archaic that nobody knows what it means. . . . There are still libraries out there with library cards and they have to go back and retype all those library cards if you change your—

Interviewer: Oh my! OK. Your words.

Librarian: So they don't change very often. . . . You just know from looking at them and knowing their vocabulary that [contacts] is not part of their subject vocabulary. Now, there may be a book out there called "The Reliability of Connector Contacts." There may exist even one book out there with that name! But there are lots of books out there that are about that. In fact, more likely it's going to be a part of a book. A chapter of a book that deals with, most likely, connectors, maybe something that deals with reliability of electronic components. So I may be looking for something that contains one of those words: connectors, contacts, or reliability. Reliability with components I might be able to pick something up on. Where I'm really going to be able to pick up a lot of material though, is in engineering databases, in articles. Because there will be lots of articles about, you know, things like what kinds of platings should be put on a contact to make it better, depending on is it a screw-in, is it a plug-in? You know, the electronic-engineering databases have all kinds of articles on that. I could get him a lot of information on that, but I can't if I follow his search as stated.

The librarian here did not simply accept the client's goal as stated but reformulated it to achieve the results that he believed the client actually wanted. Because librarians usually have a better idea than the client of what will turn up in a search, they may find it expedient to simply "do the right thing" for the client, rather than engage in further negotiation. This is how they can provide just the information the client needs with as little effort as possible on the client's part.

The negotiations we studied were subtle, nuanced, tactful, and delicate. The client who wanted all the books on PCs, Windows, and Macs and the client who requested "articles on Unix" could not simply be told that their requests were ridiculous. Nor could the requests be pared down in some standard way. Conversation between librarian and client was needed to find out which avenues for pruning the search would make sense for these particular clients in their particular situations. The ineluctable human touch was present in these encounters.

An example of the importance of the human touch was revealed in a brief but telling episode in the Apple Library. A client who had been working on a computer across from the circulation desk came and stood at the circulation desk, saying nothing and looking perturbed. The librarian immediately responded, recognizing that the client had reached a point of frustration. The librarian carefully and tactfully helped the client articulate his problems and accomplish the search. The interaction was smooth, fluid, and rapid. This client was perhaps more taciturn than most, but not atypical in his befuddlement and inability to make progress. A reference librarian can help with a client's search activity even when the client is stalled or confused.

Because access to information is a fundamental need in today's world, it must be supported to the fullest, which means a living, breathing community of helpful people at the ready. The human touch will become more, not less, important as online information resources grow and information access tools proliferate.

We emphasize the interaction skills of librarians to show how librarians provide an important contribution that is different from what software tools can do. However, we do not want to suggest that there is or should be some kind of competition between people and search tools to determine which is more helpful to information seekers. They are both helpful. The library information ecology highlights the value of these complementary contributions.

STRATEGIC EXPERTISE

There is more than information therapy going on in the librarian's cubicle. Once the librarian and client come have to a reasonable understanding of the client's information needs, the search can proceed. Now a different kind of expertise comes into play—not expertise in interaction, but technical skill and knowledge of where information lives and how it is organized. This is where the fun of detective work begins.

Searches are planned ahead, in order to provide the best results at the lowest cost. For a particular search, the librarian might use print materials, online materials, or both. A search might be a single query to an online database or it might consist of several steps in sequence, with the result of each step providing clues for the next.

Librarians have considerable technical expertise. To begin, they know a great deal about different information sources. They know which topics different sources cover, their reliability (fact, rumor, or something in between), costs, particular slants or perspectives, breadth of coverage, frequency of update (daily, weekly, monthly), and unique coverage (the only place to find information on small Swiss companies).

Choosing good sources is a powerful means of setting the search off in the right direction. Once the sources have been chosen, other technical knowledge comes into play. Librarians know about the different components of each information collection and use these to good effect. For example, in newspaper databases they can search for a term in headlines or lead paragraphs to find articles that are truly *about* a particular topic, rather than searching the full text of the newspaper and obtaining many glancing references buried deep inside articles.

Many databases have a specific vocabulary of terms and synonyms. The librarian must find the correct terms to perform successfully a search in a particular database. The right terms may *not* be the most obvious ones. For example, AIDS appears only under "immune deficiency syndrome, acquired" in one particular online database. The librarian decides in advance which search terms to try within each source, based on the vocabulary that is appropriate for the topic, geographic area, and time period of the search.

All this knowledge and experience provides the librarian with handles that can be used to move the search spotlight in one direction or another. As information resources change over time, the librarian's search strategies change along with them. So when new technologies arise, such as the Web with its variety of information and search engines, librarians adapt their strategies.

Looking at intermediate results is an important aspect of good searching, whether the searcher is a librarian or the client herself. Typically, a librarian adjusts his plan as he examines the intermediate search results, adding or discarding vocabulary terms or databases as needed. The librarian often consults the client to see if the results seem to be on target. People usually know something about the topics they are investigating, and this prior knowledge can be used to check whether a search is on track. Is there a seminal article that should have popped up but did not? Are the results unbelievably sparse? With these important kernels of information in hand, the search process can be further refined.

Librarians often take advantage of the different forms of expertise among their colleagues. At the Hewlett-Packard corporate library, one librarian specialized in business, another preferred chemistry and certain technology areas, and others had other specialties. Sometimes librarians have preferences for different kinds of searches, as well as expertise in different knowledge domains. For example, while all librarians can do patent searches and citation searches, some enjoy these types of searches more than others. Librarians talk to one another, exchange tips, and sometimes pass on a search task to someone whose expertise is a better match to the problem.

At the end of every search, the librarian packages the results before delivering them to the client. At the very least, she removes duplicate items that have come into the results somehow—there is no point in coming across the same thing twice. She might also remove obvious mismatches to the client's needs, results that technically match the search criteria but are inappropriate for the client. Her ability to weed out these "false drops," as librarians call them, depends on the depth of her understanding of the client's interests. At the client's request, the librarian might rearrange the data, add special formatting, or delete information she knows the client is not interested in seeing.

Some of the special contributions of librarians are grounded in their own histories and experience. When librarians obtain search results, they read them! In order to weed out false drops, librarians read citations, abstracts, and sometimes portions of articles or books. Reference librarians provide tremendous value to clients in returning useful, relevant information, unencumbered with extraneous material. Irrelevant material can only be cast out if it is *understood*—the search criteria alone are not enough to ensure relevancy, since of course they led to the irrelevant material popping up in the first place.

"Relevance" is not a one-dimensional concept that can be captured by a list of the client's favorite keywords. It requires judgment based on the goals and state of the client's current activity, the quality of the material itself, and knowledge of the overall quality of the source of the material.

Another special contribution of librarians is the ability to traverse the boundary between online and paper resources. A surprising number of important sources do not occur in electronic form and may not even be indexed online. Many statistical tables, for example, are not found online. Annual reports with pictures, internal corporate reports such as technical reports and white papers, and older publications, such as old or extinct journals, are not online. And yet such sources may be crucial to the work of a client. Librarians are able to locate these sources that would, in many cases, be impossible for the client to find on his own. In the future more and more material will be online, but there will always be valuable archived material that never makes it to electronic form and fragile documents that cannot be scanned.

A librarian at the Apple Library sent us the following email relating an experience she had had:

> Got a call at 10 am from Gil's [Amelio] office. He wanted a copy of the full text of a speech given by Walter Lippmann in 1939 or 1940 which had a specific line in it, which was read to me over the phone. He needed it by noon. We have a couple of good reference books in the library of famous quotations and we found the reference pretty quickly. It was a speech that Lippmann gave at Yale in 1939 to a reunion of the class of 1910. But it wasn't the whole speech and Gil wanted that. We called the Library at Yale and they said they had the speech and were going to fax it to us . . . we also went on the Internet to see if we could find it more quikly and never found it. So everything isn't on the Internet!

Librarians also bring a special contribution in their ability and willingness to provide informal evaluations of information. At Apple, librarians were sometimes heard to mutter that a particular commercial database was "lame" or "pathetic." The librarians were referring to the fact that the quality of data in a particular information database varies due to a number of factors. The large commercial databases contain millions of items, and because of their sheer size they are not always perfectly up-to-date or complete. The indexing scheme in a particular database may change, providing better or worse access depending on the changes. Databases may be set up to receive new information in a two-step scheme, where a "first load" contains only citations or pointers to the articles and a "second load" with the complete text of the articles is done at a later time.

A skilled librarian is aware of these realities of database management and can investigate further if he believes that more data may be forthcoming. The librarian can answer the questions: Are the search results too sparse, too dated, too off the mark? A client has a much harder time doing this, because she is not in daily contact with the databases and cannot venture reliable guesses about whether she should be getting better data back from a particular database. For legal and commercial reasons it would be difficult for an online service to provide the frank assessment of a particular database that a reference librarian can provide.

Again, we must emphasize that the work of librarians and search tools is complementary—the success of each depends on the other. Librarians and library clients could not succeed in their information quests without powerful tools, and the online tools can easily fall short of delivering the best results to clients unless librarians or other experienced searchers are there to help. Information needs come in many flavors and categories, and there must be a corresponding diversity in tools and services to meet different needs. Information clients will be best served if there are interconnections between tools and services, so people can move from one to another smoothly. A congenial mix of technologies and human resources marks a healthy information ecology.

BUILDING RELATIONSHIPS

We have talked so far about a triad of information seeker, librarian, and technology in the context of one-time searches. But there is more to the picture if we look at how people find information over time.

Librarians and clients in corporate libraries often work together repeatedly and get to know each other's styles. In the library's information ecology, there are opportunities for successful adaptations and mutual dependencies to develop as clients return to their favorite librarians. Many of the clients the reference librarians serve are "repeat clients"; that is, someone who has a good experience with a particular librarian is liable to request service from that librarian in the future. The librarians at Hewlett-Packard and Apple enjoy serving repeat clients, since they feel they can best serve those whose work and habits they know. For their part, clients find it appealing to have a personal librarian.

In our discussions with library clients, we learned about several situations that bring clients back to the library to continue a search. Some clients want the same search repeated on a regular basis. For example, a financial analyst might want to monitor the financial health of major competitors each quarter. Is one of the competitors doing especially well, so it should be watched with extra care? In these search situations, the client initially works with the librarian to set up and fine-tune the search, then leaves the librarian to perform the search at agreed-upon intervals. Here is a typical monitoring search, described by a financial analyst:

> Well, there is a few, probably about four, companies that I follow on a quarterly basis, so I just mark down on a calendar when I expect to hear their public earnings announcements. So I will give Joan [the librarian] a call up around the same time every quarter.

First the search has to be set up; someone has to figure out which are the best information sources and which terms and strategies should be followed to get the best results. Once the search has been set up, the librarian can carry it out unchanged until the client decides that the information out in the world has changed enough to justify an overhaul, such as a new competitor becoming a force in the marketplace.

It is easy to imagine software agents carrying out some of these repeated searches. The librarians we talked to would be happy to give up their routine searches, since the searches they enjoy most are the ones with some challenge to them. One line of ecological evolution we can imagine for the work of librarians would be for them to create well-designed software agents for others.

However, other repeat clients would still rely on the librarians' consulting expertise. Many renewed searches we observed had a more open-ended flavor, as the client explored a topic in an improvised fashion. The interesting aspects of each intermediate group of search results might suggest an entirely new focus and direction for the search. An economist gave us a flavor of her exploratory searches:

> Okay, maybe I will say I want market numbers, I want industry numbers, I want any kind of time series data or references in the article to time-series data. Or I'll say, just from the articles that you are pulling from this area, let's say environmental regulations and information technology, what are the kinds of things that you are seeing in the articles? I have no idea what's there, can you get a sense of what's happening . . . can you get a sense of what's going on? So sometimes you can use it as sort of an exploratory tool.

Clients use exploratory searches as a way to learn about a new area. A management consultant described a situation in which he needed to learn "whether what we are talking about is smaller than a bread box, bigger than a house—just size it."

The reason these open-ended searches lead to repeat business for the librarians is that one exploratory search often leads on to another, as people absorb information from the first set of results and think of new directions to pursue. Librarians sometimes sense that a search is not really over and keep an active folder on the client with information about what has been done so far. If and when the search is reactivated, this information will help the librarian set up the next step more quickly and accurately.

This kind of wandering search benefits from a human being's tolerance for ambiguity. It gives an idea of how self-service tools and expert librarians can share the workload in the library. Simple searches that are intended to answer a clear, well-stated question can be done even by relatively inexperienced walk-up users. Exploratory, open-ended, vague searches that lead to thousands of resulting books or articles usually can benefit from the help of librarians. A diversity of available resources—both human and technical—serves clients best.

In another kind of repeat search visit, the client carries out several steps in a systematic information-finding plan. For example, someone who helps put together corporate business deals described his searching this way:

> I spent the last nine months working on our acquisition of Widget Company's software business. Once again I made very extensive use of Joan . . . I got a complete dump on everything that has been said or written about Widget in the last two years . . . One area that you frequently look at in merger and acquisition work is, you want to look at what is known as comparables . . . on comparables what you are looking for is comparable transactions. . . . And then once again, once you get into comparables then you want to get more information on these transactions you see, then you are going into general business information to find out about this specific transaction, or you are going back into the financial stuff about the people that participated in each of these transactions. . . .

In this example, the merger specialist described a three-step search process. First, he got a general overview of the candidate company. Next, he tried to find comparable deals that took place recently in this industry (which is like looking at comparable real estate deals to decide on the current value of a house). Finally, he picked the most similar of the comparable deals and looked more deeply into their details.

There are direct links between each step in this search plan and the next. For example, the comparable deals that were investigated in detail in the third step above came from the second step. However, these steps could not be executed in sequence automatically, without the merger specialist's own evaluation and analysis along the way. So this kind of search is not as obvious a candidate for software agents to perform, although there is potential for hybrid contributions.

There are real rewards for library clients and librarians to work together over repeat visits and get to know each other's histories. To encourage this experience, sometimes Hewlett-Packard librarians have become liaisons to work teams, sitting in on meetings and tracking team members' activities. In this model, the librarian develops an understanding of the team's information needs and can suggest searches and establish the search context more easily.

Even when the liaison role is not official, librarians can infer a team's activities and progress from its members' repeat visits to the library over the course of a project, and they can use this information to do better searching for the team members. For example, a team might look for breadth of coverage during the early exploratory phase of a project and depth in specific areas later on. If a librarian knows the current phase of the project, she can tailor her search results.

While skill in searching online databases and paper sources is critical, the technical aspects of a librarian's job are in some sense less important than the art of doing a really good search for a particular client with a particular need. As one librarian put it, "Where the actual effort [of a search] comes in is really making sure you know what the user wants." Another librarian stated, "Thirty searches from one user is less work than thirty people asking for [a search on] one company."

In the interview fragment below, the librarian explained how he dealt with a search for clients with whom he had worked extensively in the past. In the following exchange, the librarian read the actual written search request to the interviewer and then a discussion of the search ensued.

> *Librarian:* That's Fun-and-Games which is a division that has about 30 people, but basically reports directly to Spindler. We've done a lot of work for them so I know the kinds of things they usually [want] . . . And [the client] writes,

"I'm pulling together a situation of a review for our fun-and-games business. It would be very helpful to have any recent articles, reports—last six to nine months—that discuss or describe the following: overall trends in fun-and-games, for example the trend towards better graphics, technology trends and overall sales and key competitors such as Are We Having Fun Yet, etc. International market opportunities and challenges, most important on the list are the trends and competitors. Let me know if you have any questions or if you can have this information by early next week. Thanks for your help."

OK, because we've done work for these people before, I know that they generally prefer in-depth kinds of material which, frequently on a report like this, they will have four to five people working on analyzing the request.

Interviewer: You mean analyzing what you send back to them?

Librarian: Right . . . So, I know that I can send them 100 or 200 articles and they aren't going to [throw them away]. . . .

I also know that—well, I know lots of things about these particular people from having worked with them. I know that they tend to be on a rush basis all the time. The fact that they've given me a week is very unusual. That they are very—tend to be interested in numbers; numeric data that they can put together because they were analysts and they like to be able to analyze. . . . So any numbers that I can get for them, they're going to like to see. They are going to be interested in very current stuff.

. . . I would know right off the bat that these people aren't interested in anything earlier than January. Let's see, I know that they're going to be specifically interested in each of those four points. . . . And they're going to be interested basically in general business, particularly news wire kinds of things, and other analysts. Analysts like to plagiarize other analysts.

Librarians use their historical knowledge of clients to avoid asking lots of direct questions. They seem to prefer to resort to the technique of asking the client for information only when necessary. As one librarian said,

> [T]he searchers get these repeat clients, of course, and they do get to know the people. It seems to be common, that once you know that person, you don't have to ask him as many of these questions, because you kind of know what to expect. If something is different, that will probably come to light. You can kind of say, "Well, here's what they wanted before, for the last four or five searches."

In particular with repeat clients, the librarian comes to understand the client's activity that is prompting the searches. A process benchmarking specialist described his interactions with the librarian:

> *Benchmarking Specialist:* Let's say that I decide I work with [a division] and say we need to study quick response logistics, quick response team management programs. So I say, "All right, let me go out and do a little research and find out some names that tend to come up." You have got a few ideas, who it is, and let's say you want to do some research so I will go to Joan and I'll say, "Joan, here is what we are studying, it is quick response logistics. Now let's make that a sort word. Now the problem with it it's sometimes maybe considered to be a fairly new phrase, so what could be some of the components. So look under supply chain management. Look under order fulfillment. Look under inventory management. Look at it under logistic partnerships. Look under blah blah blah." You know, all types of things like that. Usually I want fairly current data so I will typically cut it off at 1990 or something or '91 even, coming forward.
>
> *Interviewer:* And is that because these processes are new?
>
> *Benchmarking Specialist:* Yeah. . . . So that is the way it works and early on in the relationship with Joan, I will go up there and sit down with her face to face and walk through it cause I didn't know her and I really didn't have necessarily any reason to trust her. And like, all right, "Here is the deal" and she would take diligent notes and whatever and I would go tromping off. She would pull the information and send me abstracts or full articles a week later or something. Now I simply just pick up the phone and call her or leave her a voicemail message or something and say, "Here's what we need, can you tackle it. . . . "

As we can see in this exchange, the knowledge and trust in the relationship work in both directions. The benchmarking specialist had to get to know the librarian before he was confident that she would reliably come up with the information he needed.

In many corporate and government settings a great deal of proprietary material exists—the kind of thing that cannot be placed on the Internet, even if the Internet were considerably more secure than it is today. Often such material is of great value to someone in the organization who does not know it exists but who could benefit from reading it and who can be trusted to safeguard it. In the Apple Library, the librarians described cases where they brought two clients together who were then able to share valuable proprietary information. The librarians did not keep files of proprietary documents but because they knew who was doing what in the organization they could often guess where useful proprietary information was likely to be found. They then brokered a direct connection between clients who themselves established the appropriateness of sharing the information.

In another case, a librarian guessed the person (outside the company) who might be working on a patent relevant to an Apple client's needs. The librarian made discreet inquiries, found that the person was indeed working on the patent, and then allowed the Apple client to make contact with the author to obtain the patent materials (which were not ready for prime time but which could be sent to the Apple person, whom the inventor trusted). The amount of money saved when an organization does not have to duplicate effort because information is in the right hands is difficult to calculate but it is undoubtedly quite large. Again we have a situation in which tact, diplomacy, and good judgment are critical in the information finding and sharing process.

LOOKING TO THE FUTURE OF THE LIBRARY

The characteristics we find most important in the library ecology are the diversity and complementarity of technological and human resources, and the presence of librarians as a keystone species giving unique shape and strength to the library ecology.

Librarians have a flourishing information-consulting practice, but not enough people know about it. The first strategic questions to ask in an information ecology involve taking stock of the current situation—noting what can be observed, analyzed, and felt. This can be hard when a substantial piece of invisible work is involved, as is the case in the library. In this chapter, we have tried to repair a gap in awareness about information searching by describing the practices and contributions of reference librarians.

The next step for people involved in library ecologies (librarians, clients, and information tool developers) is to use this awareness in deciding what directions to take in developing new technologies and practices.

Sound suggestions come from the library community. For example, Terence Huwe, Director of the Library at the Institute for Industrial Relations at the University of California, Berkeley, suggests that librarians actively participate in the development of technologies such as digital libraries, forming strong partnerships with designers.[7] He suggests that librarians expand to new niches such as providing information services to NGO's (nongovernmental organizations), which are important forces of change in developing countries. Opportunities exist for both virtual and face-to-face collaborations in this sphere. Huwe makes the interesting suggestion that librarians reconstitute their professional organizations, recruiting new members who are information professionals but not necessarily professional librarians. Such people might include computer scientists and social scientists. Librarians can also migrate to other professional organizations for a two-way flow and improved dialogue.

With the advent of the Internet, we believe librarians are more important than ever. Librarians are information consultants who are oriented toward quality information service. We need more of this in the Internet world! Librarians are becoming actively involved in the design and provision of Internet services, and we hope this trend continues and intensifies.

There is plenty of room for more diverse and advanced information consulting practices to grow. Librarians can have a major impact on developing catalogues of Web materials—in the library tradition, there is a several-hundred-year-old body of knowledge about cataloguing to draw upon. Many are already involved in this activity online. Librarians might do more information consulting remotely for a variety of clients (as they already do in the Hewlett-Packard corporate library, which serves employees all over the world). They can and do evaluate and assess online information collections and search tools. Librarians could create canned searches, or "agents" for well-defined, repetitive searches, such as those that need to be run daily, weekly, monthly. Designing agents would be a new role for librarians but one that grows naturally out of work they are already doing. In general, we hope that librarians will continue to expand their roles, to participate regularly in conversations about the design and use of all kinds of information access tools.

There are useful and complementary relationships between information seekers, librarians, and technology in the library. The diversity of the library information ecology is exemplary in its excellent mix of people and technological resources working together well. The presence of human guides and experts in the library is crucial. As more people gain access to online information services, even more guides will be needed to help. As in the library, we believe that such guides will be keystone species wherever they are found.

8
Wolf, Batgirl, and Starlight: Finding a Real Community in a Virtual World

A researcher in a lab in California and a sixth-grade student in a Phoenix, Arizona, classroom look intently at their respective computer screens. Although they are hundreds of miles apart in the geography of the physical world, they are side by side in the "virtual" world of Pueblo. Both the researcher and the student are connected to Pueblo over the Internet, and they can hold a conversation by typing comments back and forth. The researcher has heard about this student, who is a well-known person in Pueblo, but the two have never met.

The researcher and the young student have chosen the names by which they are known in Pueblo. The researcher calls herself Tinlizzie. The student calls herself Wolf, a name she chose after doing a school project on wolves in the fourth grade. The student, who is female, has set Wolf's gender as male. Each participant creates a description of his or her online character. Wolf's description is as follows:

```
Wolf stands at 2'4. He weights about 105 lb.
Wolf has brownish-gray colored fur,and a long
1'2 bushy tail. He has brown eyes. His teeth
are sharp and pearl white. Wolf walks on four
feet, somtimes on his hind legs. He has a
strong and loud voice that can be heard from
miles away. He is very friendly and like to
talk. Wolf is 11 years old. His favorit food
is roadrunner and sheep.
```

Every participant in Pueblo also writes a brief description of his or her real-world identity. Wolf's real-world description says:

> Julie Hudson is an 11-year-old Navajo and
> Apache student at Longview Elemetary. She
> loves to play poker, speed,and spit. she also
> like to read, write, draw, ride horses at her
> grandparents house, and most of all to be with
> here with all her friends in pueblo. She has 2
> dogs Duke & Fluffy, 4 cats Bubble Gum, Tiger,
> Snow ball, and Cotton ball (little brother
> named Cotton Ball).

The researcher has wanted to meet Wolf for quite awhile. Now she types these words:

> page Wolf Hi there, can I introduce myself?
> I'm Tinlizzie. I've been on Pueblo for a long
> time and I know you have too, but I don't
> think we've ever met.

Anyone who is connected to Pueblo can "page" someone else who is also connected at the same time. What Wolf sees on his screen after Tinlizzie's page command is this:

> Tinlizzie pages, "Hi there, can I introduce
> myself? I'm Tinlizzie. I've been on Pueblo for
> a long time and I know you have too, but I
> don't think we've ever met."

What Tinlizzie sees on her own screen after she has paged Wolf is this:

> You howl your words to Wolf.

This line is provided as automatic feedback that the page command Tinlizzie typed was actually carried out—now Tinlizzie knows that her message has appeared on Wolf's screen. This kind of online feedback is helpful. In a face-to-face conversation, there are clues in facial expression and posture to tell us our words have been heard, but these are missing in text-based online conversations. Wolf has customized the feedback text that is shown to people who page him, so it goes well with his online character.

When Wolf answers, Tinlizzie sees these words on her screen:

```
Wolf howls to you.
He pages, "hi"
```

While this is not exactly a lengthy invitation to continue the conversation, it is at least an indication of some willingness to move forward. Tinlizzie now adds a friendly "physical" gesture and makes a request, by typing these words:

```
page Wolf :grins. Hobbes told me you are
making a zoo. Is there any chance I could
visit it sometime?
```

The colon punctuation (:) is a shorthand for "emoting" a feeling or a gesture. Here is what Wolf sees on her screen after Tinlizzie's page command:

```
Paging from afar, Tinlizzie grins. Hobbes told
me you are making a zoo. Is there any chance I
could visit it sometime?
```

Wolf responds cordially:

```
He pages, "yes, just walk to city zoo"
```

The virtual world is organized into separate places. Anyone who is connected to Pueblo has a current location—some particular place in the virtual world where their character can be found. By typing commands like "north," "up," and "walk to," people can move around within the local geography. There are over five thousand different locations, so the probability of stumbling across a particular destination by chance is remote. Tinlizzie follows Wolf's instruction, typing these words:

```
walk to city zoo
```

But here is the system's feedback to Tinlizzie's walk command:

```
Searching for "city zoo", please wait.
Found the CITY ZOO. Searching for a path from
here to there. Please wait.
There is no walkable path from here to the
CITY ZOO.
```

The zoo seems to be one of those places that are not connected to the main Pueblo geography. Most places are connected together through "exits" leading from each place to its nearest neighbors, but it is up to each place's creator to decide where it lies in Pueblo's geography. It is certainly possible for a place to be detached from the rest of Pueblo. Wolf seems to have set up his zoo to be inaccessible for now, though he does not remember that when he gives Tinlizzie advice on how to get there. At least, it does not seem to be connected in any way to Tinlizzie's Garden, which is where Tinlizzie is located.

So Tinlizzie asks for help:

```
page Wolf It says there's no walkable path
from where I am.
```

After a brief pause, Tinlizzie sees these words appear on her screen:

```
Wolf pulls up in the PURPLE MUSTANG with
screeching tires.
Wolf gets out of the PURPLE MUSTANG, closing
the door behind him.
```

Apparently, Wolf has come to help "in person." Tinlizzie and Wolf greet each other with waves, something they did not seem to need to do when they were just talking to each other from different Pueblo locations. To do her part, Tinlizzie types:

```
:waves.
```

As a "visible" result of her gesture, she and Wolf both see on their separate screens:

```
Tinlizzie waves.
```

Wolf responds:

```
Wolf wave
Wolf says, "enter pm and i'll take you there"
```

The word "pm" is an abbreviation for the Purple Mustang. Abbreviations are commonly used for people, places, and objects such as this car, to save time and effort for typists.

Tinlizzie follows Wolf's directions by typing:

```
enter pm
```

But she encounters a problem right away:

```
The PURPLE MUSTANG is locked and you can't get
in without the key.
```

Tinlizzie reports the problem to Wolf, who hastily unlocks the Mustang's door and asks Tinlizzie to try again (all of this by typing additional commands). Now Tinlizzie can go inside, by typing again:

```
enter pm
```

Tinlizzie sees on her screen:

```
The PURPLE MUSTANG
You sit on a beautyful almond colored leather
seat. The steering wheel has a maching leather
coat to it. The car it smells of cherrys.
The windows are down
Outside you see Tinlizzie's Garden
Wolf is out there.
Wolf hops into the PURPLE MUSTANG, closing the
door behind him.
Wolf just buckled up.
```

Following Wolf's lead, Tinlizzie buckles up too, by typing:

```
buckle seat belt
```

These lines appear on the screen:

```
Tinlizzie just buckled up.
Wolf asks: city zoo please!
Wolf pulls out a map . . .
You feel yourself thrust back in your seat as
Wolf brings the car to 50 Moo mph!
Wolf steps on the gas pedal and you take off!
All the scenery passing by has made you quite
sleepy and you doze off.
You wake up rather startled that you've
arrived already.
Wolf slams on the brakes and smiles as you
arrive.
```

And on they go. Wolf shows Tinlizzie the areas of the zoo that he has already completed and those he is still working on. He encourages

Tinlizzie to take zoo food from handy food dispensers he has made and feed it to the animals. All of Wolf's creations and all of Tinlizzie's and Wolf's speech and actions are made up of textual descriptions and commands.

Let's return to the physical world. Wolf/Julie is doing her zoo development while she is working in her sixth-grade classroom, during the time she signed up to use one of her classroom's computers. Everyone in her class—in fact, everyone in the sixth grade—is part of Pueblo. Pueblo participants also include many younger children at the same school, recent graduates, teachers, local senior citizens, students from other geographic areas, and researchers. Altogether, there are hundreds of people involved. In this chapter, we will talk about why these people are building a virtual world together and how they are contributing to an information ecology.

Most of Pueblo's participants are students and teachers at Longview Elementary, an inner-city elementary school in Phoenix, Arizona. Longview formed a collaboration with researchers in technology and education at Phoenix College and the Xerox Palo Alto Research Center (PARC) in Palo Alto, California, to set up and develop Pueblo as an experimental learning environment.[1]

Virtual worlds are more than imagined geographies; they are sites that allow people to band together and form enduring communities, with all the multiplicity of views and activities one finds in traditional communities. Vicki, also known as Tinlizzie, has been one of the participants and observers of Pueblo, as a member of the Xerox PARC research group. The illustrations and interpretations we discuss in this chapter are from Vicki's perspective of what happens there, drawing on some of the collective insights of the Pueblo community about its own practices.[2]

Pueblo has a diverse population, and all of its members can have a real impact on the activities and development of the virtual world. Diversity is crucial to the success of Pueblo. The ecology of the school-centered network community is unusual in the extent to which teachers, students, administrators, and community members combine their insights and energies to make the technology work for them, to suit their many different needs and interests. Teachers have carefully absorbed the technology into the classroom, managing the flow of activities and children as they move back and forth between the virtual world and the real world. Each teacher decides how Pueblo will fit into the school day in his or her classroom.

In addition, the built-in characteristics of the online world encourage and support grass-roots effort across the community. Anyone can talk to anyone else who happens to be connected at the same time. Anyone can build and extend the geography of the world, for others to see and share. The social and technical environments are very closely linked. The co-evolution of technical features and social practices is evident in this online world.

Our analysis of Pueblo will emphasize the application of values to technology use, diversity in the ecology, and the thoughtful coevolution of tools and social practices. Before exploring these themes in more detail, we will talk about the nature of virtual worlds and set the scene for Longview's participation in Pueblo. Virtual worlds are intricate social settings with links to the physical-world settings of their inhabitants.

VIRTUAL WORLDS

What is a virtual world? It is a computer-based environment that mimics the ordinary, physical world we live in or imagine. While visiting a virtual world, participants can communicate with one another, explore, and sometimes manipulate the objects they encounter as they look around. The simplest kind of virtual world is composed of text. This is the kind of world Pueblo is. Visiting the world feels a little like walking into a novel or an adventure game.

Participants' computers are connected over the Internet. To connect to a virtual world, participants must know the address of the computer where it is being hosted. While connected, a participant types textual commands (such as "up" or "south") to move around and sees textual descriptions of these locations scroll by on the screen.

In a text-based world, there are no graphical images, as there are in some other virtual words. Through reading and writing, inhabitants of virtual worlds can imagine quite effectively that they are immersed in these invented settings, with an awareness of the others who are there.

Each virtual world has four central characteristics that help shape the activities that can take place there. The first is *geography,* the metaphor of connected places, and the participants' role as builders of new places and things. (Remember Wolf, with her zoo, Purple Mustang, and other objects.) The second characteristic is *identity,* the freedom of participants to choose how they will present themselves to others. The third property of virtual worlds is *communication and awareness* mechanisms to suit different styles and groupings of participants. The fourth is a rooted sense of *community* that can develop for long-term inhabitants. These four characteristics are not attributes of the technology alone. They reflect the kinds of things people choose to do, not just what they can do, with virtual-world technology.

Online virtual worlds are an outgrowth of real-world fantasy role-playing games such as Dungeons and Dragons, in which players take on roles and enact adventures together. The first online multi-user virtual worlds were developed in the early 1980s.[3] Like the physical-world versions of these games that are played with paper, pencil, and dice, the online versions allow multiple players to develop their own roles and actions. Players invent their dialogue and their surroundings as they play.

Hundreds of different virtual worlds have become available on the Internet. Many have themes related to fantasy and science fiction scenarios. There are others too—Postmodern MOO for discussion of postmodern thought, BioMOO for a community of biologists, and Diversity University for online college-level courses. Like Diversity University, Pueblo is centered around education and learning. Unlike many virtual worlds, Pueblo is centered around a particular place in the ordinary, physical world. Though Pueblo has scattered network participants, it began as an extension of a community in Phoenix, Arizona.

LONGVIEW ELEMENTARY

At Longview Elementary School, a kindergarten through sixth-grade school with a large population of Latino and Native American students, over half of the students are learning to be proficient in English. Nearly all of these students are poor enough to qualify for free or reduced-price lunch at school. The children are sometimes exposed to violence and substance abuse in their neighborhoods. The school itself is a safe haven for the students. Its doors are open from early in the morning until early evening, so students have a place to be if they don't have adult supervision at home.

Here is how one teacher describes the particular needs of a new class in the fall, just after students have returned from a Pueblo orientation field trip:

> [S]everal of the LD [Learning Disabled] kids are in my room. I think almost every kid has a special need of one sort or another, many have more than one. . . .
>
> Several of the kids are ESL [English as a Second Language] and are just now transitioning into English. Several are still adjusting to school after a summer on the streets . . . Three of the kids in my class are on meds . . . there is one kid whose family is being evicted. There is at least one known case of abuse. One kid worries about his mom being beaten by her boyfriend. There are the single parent families, of course. Gang activity. On and on.
>
> So, I don't know what else to say except that all of them, except [one child], are quite excited about Pueblo, and seem to need that paging capability like some of us need oxygen—another indication, I think, of their extreme need for attention and communication.

From the perspective of the Longview teachers and principal, a virtual world offers an experience that can compensate for some of the disadvantages in these students' lives. It brings children into regular contact with adults who can give them personal attention.

Since 1993, teachers and students at Longview have been experimenting with the use of virtual-world technology in the classroom, using refurbished computers donated by local companies. (It has been fortunate that the virtual world exists only in text, because the earliest donated computers were of a vintage that predated windows, mice, and graphics.) Longview was invited to participate in the online world by Phoenix College education faculty members Billie Hughes and Jim Walters.

Hughes's and Walters's long-term goal has been to develop a "K-Gray" cross-generation learning program at Phoenix College, with the network community as a centerpiece. The Phoenix area is a popular place to retire, and there is a large population of active seniors who are interested in continuing their own learning through courses at senior centers and community colleges. Phoenix College educators do not want to serve only those who make their way to college on their own. They want to reach out and actively encourage people throughout the local community to see college as a possibility for themselves—whatever their age. As Walters puts it, "A college education begins in kindergarten."

Before partnering with Longview, Hughes and Walters had experimented with an earlier virtual world in the college setting.[4] They wanted to learn what a virtual world environment was like and to see whether students and faculty would use it as an opportunity to explore new ways of teaching and learning. They were encouraged by the results. With Phoenix College's orientation to community outreach, extending beyond college students to include younger and older participants was a natural step. In their vision of a cross-generation learning community, all participants will learn and contribute to the learning of others.

For Longview, developing a relationships with Phoenix College was a natural fit to the school's existing programs. Jo Talazus, Longview's principal, explains how Pueblo fits into her larger goals:

> I am always searching for the ways and means to create a community school—reaching out to the local and worldwide communities. I firmly believe that the greater the number of successful adult relationships a child establishes, the greater the likelihood of success for that child in the future . . . . Therefore, our school has business tutors, police mentors, classroom grandparents, Hispanic attorney mentors, Junior Achievement volunteers, Phoenix College partners, and others. All but Junior Achievement provide one-on-one contact between adult and child.

To make it possible for Longview to participate in the K-Gray program, Walters placed ads in the local newspaper to solicit computer donations. Because the donated machines were old and idiosyncratic, they were not always reliable. In a classroom with five or ten computers, several were likely to be acting up at any particular time. Hardware (especially donated hardware) needs regular, personal attention in a school setting. In the early days, Walters often patrolled the halls at Longview to make sure the computers were holding up. After a time, the school was able to hire a technology aide to monitor and solve equipment problems, though this remains a challenging logistical and financial issue.

With the scarcity of funds, teachers had no paid release time during the school year to spend on new curriculum planning and design. But what they lacked in money they more than made up for in energy. In summer camps, children and teachers worked and played together online, practicing reading and writing skills in a highly motivating, constructive environment. At the end of one successful summer camp, Walters persuaded Vice President Al Gore to deliver a commencement address—by written proclamation rather than physical presence, but still an exciting event that showed students and teachers the potential of building a network community.

Later, Longview and Phoenix College were joined by Xerox PARC researchers. The additional grant funding that came through this partnership helped provide teachers with release time throughout the school year, summer camp support, and some new equipment.

PUEBLO AS AN INFORMATION ECOLOGY

One of the most important aspects of the Longview-Pueblo information ecology is that the technology-based virtual world has become an extension of the real-world classroom, inheriting its social values and educational mission from the real-life surroundings.

One popular conception of online virtual worlds is that they are addictive and unhealthy. MIT researcher Sherry Turkle and others have recounted stories of college undergraduates who spend most of their waking hours connected to a virtual world, foregoing classes and other face-to-face social encounters.[5] This kind of engagement with online virtual worlds may be found in college settings, but not in Pueblo.

Here, the virtual world is not an escape from reality—instead, it is grounded in reality. It ties together people who often already know each other and spend time together in the physical world of the classroom. Students and adults are accountable for their social behavior and their learning online, both to their fellow community members and to the school setting in which Pueblo is embedded. Values migrate from one setting to the other.

To anchor the virtual world in the real-world school, the educators in Phoenix have introduced familiar classroom practices into Pueblo.

Students have rights and responsibilities in Pueblo that are directly reflected from the school setting. They discuss their rights and responsibilities before they attend their first Pueblo orientation session. These include taking responsibility for one's own learning, paying attention to quality, promoting a safe learning environment, and seeking and providing help when needed. A list of rights and responsibilities is posted in the library. Values are given center-stage attention, both in the online and classroom settings.

Pueblo is not just a software application—it is a social setting that feels like a warm, supportive, living community. The community is bound together by the social goals, values, and expectations of its members, and these in turn have a strong and explicit effect on how technology evolves in this ecology.

CENTERING ON LEARNING

The most basic educational concern of teachers and parents at Longview is that children stay in school. Dropout and absentee rates are alarmingly high. The transience of the population at Longview is also a challenge—sometimes when children leave Longview they do not go to another school but take time out while their families are in transition, moving from one place to another. For example, by early October 1996 there had already been a turnover of over two hundred Longview students since the start of school in September. Keeping up with schoolwork is hard under these circumstances.

Children enjoy being part of Pueblo, and teachers believe that access to the virtual world is helping some of them stay in school. Teachers find that the combination of social interaction, construction, and exploration appeals to their full range of students: high and low achievers, English and non-English speakers, girls and boys, students who function well in classroom environments and students who have "behavior problems."

Students sign up at the beginning of each day for online time in Pueblo. In most classrooms in which Pueblo is used, students choose to be online every day or two. The library sign-up times for before and after school and lunchtime slots are usually full until the school closes its doors in the evening. Some teachers have volunteered their time after school in response to students' requests for access to classroom computers.

Another basic concern for this population of children is exposure to a wider world, through the involvement of adults outside the school setting. Students like talking to people who live far away and who have different kinds of experiences and backgrounds. One of the first things students want to know about a new conversation partner is where they live (followed closely by asking how old they are!). Enlarging students' horizons has been an explicit goal for Longview's principal and for Hughes and Walters at Phoenix College from the beginning. They hope that as students interact informally, on equal footing, with people of different ages, backgrounds, and experiences, they will be able to imagine new futures for themselves, including going to college.

There have been three primary curriculum links between the traditional elementary school classroom and Pueblo: language arts (basic literacy and beyond), social interaction skills, and science and math reasoning and inquiry. (We focus only on language arts and social interaction skills in this chapter, since science and math activities in Pueblo are still new and evolving.)

Social skills are an explicit part of the learning agenda at the elementary school level. Students are explicitly coached in skills such as collaborating on a task with others, resolving conflicts peacefully, and providing and accepting critical feedback. The virtual world provides a practice ground for social development.

No matter what students are doing in Pueblo (moving around, driving virtual cars, building homes, or just chatting), they are practicing their reading and writing. Students can create (and own) homes, pets, cars, food, and whatever else they can imagine and describe.

One especially motivating arena for writing and revising descriptive text is creating the geography of Pueblo. Pueblo includes a desert, mountain range, dolphin lagoon, city park, forest, and many other "rooms" (separate locations in the virtual world), all connected together in an untidy sprawl. There is no architectural review committee in Pueblo, so people can invent new places as they see fit, without concerning themselves with real-world common sense. A comfortable living room can be attached to a mountain peak. It would be impossible to draw a map of Pueblo on a piece of paper—it is not a flat and well-ordered realm, but a collage of fanciful creations.

Each "place" is created through narrative descriptions. Any participant of the virtual world can make a new place, choosing where to add to the existing geography. Let's tour part of the Pueblo home of Starlight, a sixth-grader:

```
Starlight's Palace

There are many beautiful flowers of many
colors. There are a few roses and some lilies
too. Some yellow and white dasies are growing
out of the shadows and reaching out to the
sun. The grass is a bright green and it gives
you the sensation of being very cool and
fresh. The castle is build by rocks and it has
some very tall towers. On top of the towers
there are small red flags. There is a huge
front door, you get curious to see what is
inside so you decide to go in.
    Obvious exits: Batgirl's Neighborhood . . .
<BN> and courtyard . . . <co>
```

When we type the abbreviation "co" to move to courtyard, this text is printed on the screen to announce our arrival:

```
The Courtyard

When you open the door you can feel the warm
feeling and feel very cosy. In the middle of
the room there is a large table. It is
decorated with many different flower
decorations. On the walls there are many small
flags or more likely like banners. On the
banners there are pictures of warriors with
swords and spears. It looks like some kind of
war or a battle. Then something cought your
eye on the left side of the courtyard there is
a door probably from an oak tree and then you
hear a soft melody. It is coming from the door
that you saw on the left side and you get
really curious to see from where it's coming
```

```
so you walk to the door. When you to the door
you can hear the tapping of your shoes when
you take each step that sounds like this
"tip,top,tip,top,tip,top." Then when you
finaly get to the other side of the courtyard
you open the door and see. . . . . . .
Starlight is here, asleep.
Obvious exits: Starlight's palace . . . <Sp>
and stairway . . . <st>
```

This is writing for a real audience. Starlight has taken care to engage the senses of people who explore her areas, referring to sounds, light, color, touch, and movement. She lures the reader onward, introducing some suspense at the end of each description.

Donna Spano, a fifth-grade Longview teacher, wrote:[6]

> The aesthetics are addressed [in Pueblo] in a more indirect way, by encouraging participants to describe objects they have built in ways that make the reader "see" a picture of the object in their minds. In other words, a skillfully constructed description is a work of art and a thing of value on Pueblo. This is consistent with the emphasis on quality that is the expectation at our school.

When participants create descriptions like the ones Starlight wrote for her palace, they add to the ambiance of the virtual world. Participants are motivated to write well because of the enjoyment they give to themselves and others—they can create experiences for others who visit their creations.

As Jo Talazus, Longview's principal, says, "This is a far cry from 'put your papers in the basket on the desks.'" It is vastly different from preparing an assignment for a teacher (an audience of one) who is expected to grade the work, rather than simply enjoy it. In this way, teachers are able to accomplish an educational goal through the technology that they are not as easily able to accomplish through regular classroom work as it is traditionally structured.

Students have created marvelous things, some with special meaning in their personal lives. A student whose pet cat had died over the winter vacation created a Pueblo cat with the same name and description on the day he returned to school. Here is Sarah the cat's description:

> A very smart cat who looks very good. This cat
> is in honor of my late Black-cat named Sarah,
> who died on Jan. 2, 1996. She was almost 16
> years old in human years(around 120 in cat
> years), and she could sit when we told her to.
> This cat will be as nice as the real life one
> was.

Another student with a large set of parents and stepparents created objects representing all of her family, with each relationship carefully spelled out. Another student created police officers, explaining that they helped him feel safe.

Students find the activity of creation intrinsically motivating. When one student was asked to explain what he liked about Pueblo, he said, "You get to drive cars and program things and create things you could possibly never have in the real world." Cars and pets are especially popular items for students to create, personalize, and own. As with homes, the only way to make these objects really interesting is to give them unique names and descriptions, which will be seen by others as the car or pet owner travels around Pueblo.

Here are some car names and descriptions written by elementary school students:

> Panther
> A road burnin', lucky streak

> tortoise
> The seats are as soft as hamster fur. It can
> fit 7 people. It is very luxurious and
> spacious. The windows are tinted and
> everything is automatic.

> Dodge Viper
> A sleek, happy Dodge Viper. . . . . . . It
> looks like canned tomatoes.

> roller coaster car
> Your going to love this car becuase if you
> don't hold on the ride might throw you off.
> Always remember to buckle up you will need it.

The inside of the car is lined with leather to
comfort you in the ride. There is a couple
bars for you to hold on to. Also if you really
need a food vendor for before and after ride
snacks.So hold on and press the button when
your ready to go.

Pets in Pueblo include:

T*I*F*F*A*N*Y
Jet black fur and bright golden eyes. This
kitten will whap at anything that moves.

Blasty
Slender and golden like a small mountian lion.
He will climb into anything unless he's
getting a tummy rub.

Spike
A small red dragon (about the size of a house
cat). You can FLY him.

Albert
A large, handsome, twenty pound cat who is
friendly once he gets to know you. Watch out
though, he doesn't like strangers.

Formal, documentary assessment of reading and writing gains is a
concern to teachers, even when they are convinced that participation in
Pueblo is helping their students with literacy skills. Much of the writing
students do in Pueblo follows their own interests, though teachers can
and do suggest particular ways for students to write in Pueblo.

The school district requires that students complete writing assignments
each year in several genres, such as personal narratives, poems, and
letters. Teachers face a question of how to make students' writing in
Pueblo "count," since writing takes a different form there than the
writing tasks presented in most classroom settings.

This is a familiar double-edged sword for anyone who is engaged in
education experimentation or reform. Standardized tests are a major force
in determining what is valued and taught in the classroom. That is part
of their purpose, in fact—to ensure that students are held accountable to

consistent standards. The challenge for teachers who want to experiment with new methods of teaching and learning is to find ways to meet current standards, even when the problems students are solving lead them in new directions of curriculum content and thought.

Longview's school district has taken the unorthodox position of agreeing to allow electronic text written in Pueblo to be acceptable in meeting district writing requirements, as long as the text can be printed for the district's records. Students still take standardized tests as required by the district. Although the results of these tests have not yet shown any clear relationships between the use of Pueblo and improved test scores, teachers and administrators are continuing to explore ways to understand the impact of Pueblo in measurable terms.

INDEPENDENT LEARNERS

One of the striking attributes of life in Pueblo from an adult's perspective is the degree of autonomy students have online. For one thing, everyone can talk at once in Pueblo. In fact, each person can be holding *several* conversations at once. As typing speeds and familiarity with the environment grow, some of the older students learn to juggle several conversations at a time with enjoyment and ease. From moment to moment, students can choose whom to approach, whom to answer, where to go, and what to build. Over time, they can decide who they are (how they will look to others) and how they will go about making a unique, personal impact in the virtual world.

This degree of autonomy can be both a challenge and a benefit to teaching and learning practices. When teachers choose any kind of technology for a classroom setting, they must look at how its affordances match the ways they teach and the ways they want their students to learn. As we have discussed, technologies are not neutral but point people in certain directions. Donna Spano, the fifth-grade teacher, wrote these reflections about technology in the classroom:[7]

> Several years ago, when our district first began to focus on technology, I was part of a committee that visited other sites which had begun implementing technology into their curricula. One place seemed particularly interesting because it had adopted a packaged

computer curriculum in its K-6 system and I made the two hour trip with anticipation and great expectations.

In every classroom we visited, students were busily engaged in working on whatever program was scheduled for that time or class, and teachers extolled the virtues of the program by elating over decreased prep time and increased student on-task time. It all sounded good, and even looked impressive, but when I took a close look at what students were doing, I thought, "electronic worksheets!" Even the writing was packaged, giving students specific topics for various writing genres with step by step, fill-in-the-blank instructions on "creating" a finished piece. The teachers' role was monitoring behavior and recording progress, progress being defined by the completion of consecutive assignments. I left that place feeling like the old song, "Is That All There Is?". I understand the students' involvement in their tasks; computers are great motivators. I even understand the teachers' enthusiasm for a lightened work load; who wouldn't rather spend less prep time? But I wondered at the educational value of this system; what is it really teaching? Will the students be as motivated when the newness wears off?

This prepackaged approach to software or some variation of it is, I think, what is commonly used. . . . I don't know what success they have had with it, I think it largely depends on how you define success. I just knew when I left that visitation site that I would not feel like a success, nor would I believe that I was teaching students how to be successful if my district used a program similar to the one I had just seen. . . .

Using Pueblo for the curriculum requires some adjustments by the teacher. The teacher must be willing to give up some amount of control and allow students some freedom in choosing topics they may want to study in depth, and also allow students to decide how best to represent their learning in Pueblo. . . .

Independence in a virtual world takes several forms. Participants have many choices in how they create an identity, how they construct new places and things, and how they communicate with one another.

Independence in Identity

One of the most basic building blocks of independence is the ability to construct identity. This is an intriguing and much-discussed aspect of participation in a virtual world, which has special implications for a school-centered environment.

In a virtual world, participants can decide how to present themselves. As the famous cartoon says, on the Internet no one knows you're a dog. In a virtual world, you might actually choose to *be* a dog. Presentation of a person online comes through the character's name, description, and speech and gestures. In *Life on the Screen,* Sherry Turkle has described engagement in social virtual worlds as a kind of "identity workshop," a way of experimenting with different identities. That aspect of virtual-world involvement appears to be just as real and compelling to children as it is to the adults Turkle interviewed in her research. Here are some example names and self-descriptions students have developed in Pueblo:

```
Lana, 6th grade:
I make magic. I am a spirit of the universe.
My hair is long and purple and blue.
My eyes are blue too.
I am 5'5".
I am a slender mystical figure.
My lips are small and red.
My nose is just enough to fit my face.
Lana is only 100,005 years old in spirit but
16 in human years.
Lana is a kind, loving spirit who loves
adventure.

Snake, 3rd grade:
A nice human with a black tuxedo. I have a
nice house and the biggest swimming pool in
pueblo.

Sizzle, 6th grade:
Sizzle is a very nice person who will help you
if you need it. His brown eyes always give you
a mysterious but helpful look. He really likes
to talk and build on to Pueblo. What he likes
even more is helping people who need it. He
also likes to make friends.
```

```
Colin, 1st grade:
I am 7 years old and I go to first grade. I
know to add I know to count. I look brown and
I got black hair and brown eyes. I like to
play baseball. I know my abc. I know how to
read. I am good at math.
```

```
Hawk, 6th grade:
My character is a hawk. He likes to fly very
high. People respect me, because I am strong
and independent. When I fly high I could see
down. I have pretty eyes that are green and
black. When I am hungry I like to go and find
food to eat. I like to play with baby birds
because they are very cute. I have lots of
friend birds in my country. In my country
blows lots of cool air.
```

Unlike the online virtual worlds Turkle has studied, Pueblo has a policy that players are not anonymous. In addition to their character descriptions, participants are asked to develop a description of their real-world selves. This is still a self-presentation, not something created by others. But it is another way that teachers have attended to the information ecology by applying the school's values and expectations in the virtual world. They feel that students should always be able to find out who their conversation partners are, just as students should be able to decide whether to leave a conversation that makes them uncomfortable. Besides, these real-life descriptions are good conversation starters; they're intended to help people find others with common or uncommon interests.

When the social values of a familiar setting are extended to the use of a new technology, they are subject to new interpretations. There can be no simple translation of rules and policies into these new technical settings, because the new technologies create new possibilities for knowing and doing. When the mechanisms change, new protocols are needed. For example, what is the virtual-world equivalent of raising your hand for help or leaving a door partly closed to indicate a wish to concentrate? In a healthy information ecology, people evolve new social practices that are in alignment with the values and purposes behind their activities. In Pueblo, the values of the classroom setting are the basis for evolving online practices.

In the area of identity, teachers and students had to decide how to provide the right balance of disclosure of real-world identity and the development of an imaginative character. Having two parallel presentations has worked well.

In many virtual worlds, participants never meet in real life. But in a virtual world centered around a school or other institution, there is constant interplay between the virtual and real worlds. Most of Vicki's interactions with the children were online, since she lives in California and the children were in Arizona. But she visited Longview regularly to attend meetings and spend time in the classrooms. It often seemed that her online persona was key to the children's acceptance of her presence. On one occasion, she walked into a sixth-grade classroom unannounced, after an absence of four or five weeks. She was only three or four steps into the room when a young student approached and said, "Hey Tinlizzie, can you help me with my helicopter now?" It was as if she and the student were picking up a conversation that had only just paused—which, in a way, was exactly right.

On another visit to a fifth-grade classroom, Vicki approached a cluster of shy Native American girls, sitting close to one another, but each at her own keyboard. As she asked what they were doing in Pueblo, they did not make eye contact and gave monosyllabic answers to her conversation openers. She noticed an unoccupied computer and decided to sit down and visit Pueblo instead. As she began her session, there were some giggles and glances from the group of girls, and pretty soon one of them paged her with a question: "Who are you in real life?" She paged back: "I'm Vicki, and I'm sitting right next to you!" The girls quickly became chatty and friendly online, asking what she was doing and why she was there. When Vicki was booted off her computer by a student whose turn it was to use it, the girls came over and talked with her face to face about where she was from and how she liked Pueblo.

Children enjoy the identity play. It seems to free them from some of the ordinary standards of the school setting that link authority to the control of interactions. In Pueblo, they have permission to start conversations with teachers or other adults and to stop conversations when they are no longer interested.

In addition, teachers enjoy being Batgirl, Cleopatra, and Raptor for a change, instead of Ms. Olson, Ms. Melton, and Mr. Lightfoot. The identity shift allows a style of teacher-student collaboration and role reversal (students coaching teachers, for example) that is not so easy in the physical classroom. A Longview first-grade teacher was both bemused and delighted to see her students decline her offers of help in Pueblo— instead, they made a beeline for six-year-old Stephanie, who had quite a bit more expertise making pets and homes online than her teacher did.

Some of the most recent immigrants to Pueblo are a group of about a hundred senior citizens from the Phoenix area. The "grays," as they are called, also enjoy the ability to create their own self-presentations. Their earliest interactions with children were only online. When a summer camp was planned, the grays were invited to attend in person. Though they were excited to be part of camp, most were also very worried that the students would not accept them, once they had met them face to face and seen how old they were. On the first day of camp, the grays held back, staying together at computers in the library instead of moving out into the classrooms where the students were. By the second day they had broken the ice and were relieved to find that the students were as comfortable with their physical selves as with their online selves.

Self-presentations are only part of identity in the network community, just as physical appearance is only part of a real-world identity. Another aspect of identity is based on contributions to the community. In the Pueblo learning community, people want to make a difference, to "count," as one seventh-grader said.

Because all the online activities inside Pueblo are mediated through technology, technical skill is particularly visible and valued. Those who can help others are important people. And because helping behaviors (such as listening, problem solving, and explaining) are also learning behaviors, teachers and other Pueblo developers have devised ways of promoting and recognizing helpfulness. Pre-teens and teens in Pueblo have been particularly eager to follow these opportunities. They have volunteered to write documentation and do interactive help-desk style consulting for new learners who encounter problems.

These student contributions are similar in some ways to the jobs students take on in some physical classroom settings, such as cleaning up. The emphasis on personal responsibility and caretaking in one's own environment are the same. An extra dimension online is that the students are not just helping the teacher or other classroom authority—they are helping their peers and the community as a whole. As Vicki worked on writing new documentation alongside students, she was impressed by their dedication and enthusiasm. Students get to choose how they want to make an impact on community life, and they can meet and interact with the people they are helping.

Independence in Building the World

For reasons that are both technical and social, not everyone can be a builder in some virtual worlds. One of the key policy decisions of Pueblo was to encourage everyone to change the world, to make new places and objects there. If students could not design and build, much of their motivation to participate would disappear.

In any information ecology, broad access to and influence on the technology itself is a condition of its successful adaptation and evolution over time. To have influence on the development of technology, people need to engage with it and develop an understanding of how the technology fits—or should fit—into their own activities of work and learning.

In Pueblo, most students engage in both conversation and building. Conversation alone would be likely to leave them bored. They want to get their hands into the material of the world and leave a personal mark.

Giving students the ability to build and extend the virtual world is a good way to capitalize on their talents. Designers of virtual worlds in which participants do *not* have the capability of extending the environment have discovered that it is not easy to stay ahead of participants, generating enough new places and features to keep them entertained. For example, in Habitat, an early graphical virtual world, designers once spent many weeks creating an online adventure for people to experience, only to find that some solved the puzzles easily and proceeded to "finish" the adventure in just a few hours—to the dismay of the designers who had expected to keep people engaged in it for weeks.[8]

In an information ecology, influence comes through participation. Although students are not part of all decision-making forums, they have a profound effect on development. How they build, what they build, and the relationships they engage in all help shape community life as a whole. This is one of the most basic ideas of information ecologies. Through broad participation, different people in an ecology give it form and meaning. As we have argued earlier, people who use technology should not simply see themselves as consumers who can choose between this version or that one. They can be authors of their environments, as in a buildable community setting like the one we are describing here.

Independence in Communication

Communication with others who are online at the same time is the backbone of membership and participation in a virtual world. The text-based commands for communication are the most important for newcomers to learn first. If you are able to communicate, it is easy to ask for help with anything else you need to know.

Communication possibilities in a virtual world are more extensive than in a chat room on the Internet. (Chat rooms are online places for text-based conversation, often organized around particular interest groups or topics.) A Pueblo inhabitant can say something to others who are in the same room, as in Internet chat rooms. In Pueblo one can also page someone who is in a different virtual room, "emote" to express actions or feelings, or participate in a discussion between people in scattered locations. Another more indirect form of communication is through actions on objects. If I pick up a Frisbee, this action is reported to others who are present in the room where the Frisbee and I are currently located. Picking up a Frisbee in a virtual world is an implicit invitation to a game, just as it probably would be in real life.

The emote command gives Pueblo interactions much of their playfulness and emotional grounding. Here are some examples that show the range of possibilities:

```
> emote waves as she beams away.
Cat waves as she beams away.
```

```
> emote looks at the timeline and sees we are
behind.
Snapper looks at the timeline and sees we are
behind.
> emote laughs and laughs.
Squirrel laughs and laughs.
```

Notice that when people use emote, they express their gestures and emotions in the third person, rather than the first person (not "I smile" but "Tinlizzie smiles"). This is an unusual communication modality—in face-to-face communication we cannot easily step aside from our physical selves to offer play-by-play commentary on what we are thinking and feeling. Through emotes, people can say things about themselves that they might not wish to say *as* themselves—yet they are still things that the speakers choose to express. As Vicki has grown more accustomed to communication online, she has become increasingly attached to emoting. Though she could simply reply to a joke with a comment along the lines of "That's really funny," she finds it much more satisfying to chuckle, giggle, or roar with laughter, as the occasion demands. And rather than say "I don't understand that," she prefers to scratch her head and look puzzled.

Emoted actions have impact. One of the most recent new groups of immigrants to Pueblo are nursing home residents. Rose Pfefferbaum, who heads the gerontology program at Phoenix College, points out that very old people, especially those living in group homes, are rarely touched except as part of caretaking actions. In the virtual world, a hug (as in "Cora hugs you") has an emotional effect that is surprisingly like a real hug. The action has both immediacy and the suggestive power of narrative, which we are accustomed to respond to as experienced readers of novels and stories. Pfefferbaum believes that virtual-world participation may offer nursing home residents a level of intimacy in interactions and relationships that is absent from their everyday lives.

Choice is a key property of communication in Pueblo, as it is for identity creation and construction. All participants can choose when and with whom to communicate, using whatever style of approach or response is most comfortable. A student who is looking for conversation might check to see who is online (using the "who" command) and page someone who looks interesting or walk to a room where others are located and wait to be noticed and brought into the conversation. This is an area where traditional classroom standards have been very much relaxed. In the virtual world, it is not only permissible, but positively encouraged, for students to talk "in class."

In a world of textual conversations, some new ethical questions are raised. Is it permissible to pretend to be someone else, for example by renaming your character to have a name identical to someone else's except for a capital letter or two? One Longview student tried this (once), choosing the character name of a teacher to copy. This student had to take a break from participating in Pueblo as a consequence.

A thorny policy decision for Pueblo is how to handle situations where someone is speaking offensively to someone else—this is not an everyday occurrence, but it is apt to happen from time to time in any online communication medium. For historical reasons, Pueblo has a built-in command to allow one person to "gag" someone else—to filter out whatever that person might be saying. But is this the best way for students to handle conflict? Does the technical environment support the objectives of teachers and students here, or is it providing a temporary solution that actually gets in the way of a more constructive approach? Teachers decided that gagging wasn't a good way to resolve conflicts, and the gag command was blocked in Pueblo. Some of the researchers felt the command ought to be available, even if its use were discouraged, but the teachers' views prevailed. Each community has to tailor its own responses to ethical issues based on its own local values.

Another new policy area is raised by the question of whether it is acceptable to show others the written transcript of a private online conversation. Should teachers be able to routinely review students' conversations online, just as they routinely review students' written assignments? The technical communication capabilities of the virtual world are different from those of face-to-face communication. For example, online conversation leaves a trace in the form of words on the screen, which can be copied and saved into a file. We usually don't have to worry about people saving and replaying our words in ordinary conversations, since they disappear as soon as they are spoken. But what if everyone carried tape recorders that were turned on, all the time? We don't yet have social agreements and conventions to handle these new capabilities; we have to make them up as we go. In Pueblo, there is a guideline for all participants to keep any transcripts they make for their own use unless they have explicit permission from all participants to share them.

The virtual-world environment is highly programmable. "Spoofing" (pretending to be someone else), eavesdropping, and "gagging" (filtering out) the speech of another character can be accomplished easily. In an environment where independence is valued, there will always be situations in which people test the boundaries of acceptable behavior. The solution for Pueblo has not been to make these undesirable behaviors technically more difficult, but to work on developing shared values and expectations of what is appropriate.

Because Pueblo is a school-centered community, teachers and administrators examine questions of appropriate behavior in light of their social and educational goals for children. Their tactic has been to lead the children in a process of talking about their own rights and responsibilities, linking the two together. If you believe you have a right to privacy, then you have a responsibility to respect the privacy of others. In general, teachers have not focused on the use of technical mechanisms to prevent this or that social misdemeanor from occurring. Technical mechanisms can always be gotten around by other technical mechanisms, and this kind of puzzle solving can be appealing to young people. Instead, for the most part the social goals are left to stand on their own.

COMMUNITY DEVELOPMENT

An important strand of the Pueblo information ecology is its rootedness in community, in both the virtual and physical worlds. Membership in Pueblo is grounded in a shared commitment to learning and a consistent, explicit emphasis on values. The geographical centering on children and seniors in Phoenix, Arizona, adds another dimension to community. Many participants in Pueblo can meet each other face to face, so relationships develop through a combination of real-world and online interactions. Pueblo is neither strictly virtual nor strictly physical—it is a new kind of hybrid place, stretching across both physical and virtual locales.[9]

Children and adults share common activities online that matter to both. What happens in Pueblo is not just adults volunteering to teach or assist children. As we described in the example of first-grader Stephanie, at times students may be in a better position to help than adults. The point isn't who knows more—the point is what people undertake to do together, each making their own contributions.

It can be unsettling, though rewarding, for adults to adjust to a partnership style of interaction with students. As the grays come into Pueblo, some find it hard to understand what their roles are to be. It is both like and unlike serving as a classroom volunteer. The grays have valuable life experience and problem-solving skills to share, but they are also fellow learners. Together with the students, they are inventing a new kind of community membership as they form ties in Pueblo.

Over time, every virtual world develops its own identity. It has a particular feel to visitors and residents—it might be friendly or unfriendly to newcomers, low-key and quiet or stimulating and noisy. Several of the sixth-grade teachers at Longview asked their students to write about the meaning of community and to consider whether Pueblo was a community. Their responses capture some of the profile of Pueblo:

> Amat: A community is a group of people who
> live together in a neighborhood. Pueblo is a
> community because people build homes there and
> live there in Virtual Reality.

> Quinn: People working and helping each other
> makes Pueblo a community . . . it's fun and
> it's like our own little world.

> Marbles: I think a community is a group of
> people that help each other out and do things
> together and that think of new ideas to
> improve themselves. To me what makes Pueblo a
> community is that people get to make new
> friends when they page each other.

> Kitten: A community is when a lot of nice
> people come together and make fun games for us
> and we write to our friends. When people like

```
Jim and Hobbes work with the kids that are on.
And us too, we make houses that you can go in
and look, and cars that you can ride in with
us . . . You can do a lot like talk to people
that live in New York and people that are
older than you.
```

The friendly atmosphere and physical-world metaphor, especially in the development of homes and neighborhoods, are pointers to community feeling for these students. Both of these characteristics are emphasized in Pueblo. When teachers bring in a new class, they encourage students to build homes as soon as they have described themselves and mastered the basics of communication and navigation.

Kimberly Bobrow (aka Hobbes), a consultant to Xerox PARC and Phoenix College and a community member who is online much of the time, writes about how she tries to make Pueblo a hospitable place:

> Cynde [one of the teachers] and I feel so strongly about the idea of welcoming members to the community that we created an online Welcome Wagon, complete with brownies in a foil-covered plate. We keep a close eye on who logs in, and we are both on regularly enough to know instantly when someone new shows up. When that happens, we both hightail it to the wagon, step on the virtual gas, and zip off to the Visitor's Center to spread some of that Pueblo feeling.

The persistence of the virtual world is also important to developing a sense of community. When students connect to Pueblo, they can count on both continuity and difference. Their own places and things will still be there, unchanged from when they last departed. At least some of the currently connected characters should be familiar faces. But the world as a whole will be different. There might be fresh news in the newspaper, a new black Cadillac cruising around, a new home building site in the Pueblo Mountains, or new characters exploring the Visitor's Center. These continual changes keep the virtual world lively and interesting.

There is a balance between an organic growth approach to community development (let's see what happens) and a planned, design-oriented approach (let's get something specific done). Teachers sometimes purposefully design learning activities and sometimes take advantage of serendipitous opportunities that present themselves.

To steer Pueblo in directions that will be useful in the classroom, teachers have asked the more technically proficient Pueblo participants to implement specific features—this is the planned, design-oriented approach. For example, they requested online "plus-delta" rooms where students can gather to evaluate an activity by naming its "pluses" (things that went well) and "deltas" (things they would like to change next time). The plus-delta activity is familiar from the classroom, where students write their pluses and deltas on sticky notes and place them on a large poster in the front of the room. A student volunteer then reads these (anonymous) notes and the class discusses the feedback and evaluates follow-up actions.

Not surprisingly, the online version has some different affordances that change the dynamic of the plus-delta activity. One obvious change is that you don't have to walk to the front of the room to post a note. Instead, you just type your plus or delta and it is automatically recorded in an online list. Teachers have noticed much broader participation (especially among the shyer students) in the online version. Even when a new technical feature of Pueblo is modeled on familiar classroom activities, unplanned effects emerge in the online translation.

Students often participate in activities and develop features that are not planned or designed by teachers—this is the organic growth approach to community development. For example, an adult created a City Park. A fourth-grade student was inspired to add a lake with fish. Two teachers then provided fishing poles, with which you could (sometimes, if you were lucky) catch fish. Each person started with the material that was there and designed something new with it. Favorite online experiences and activities coevolve along with the tools and objects available in the virtual world.

At a larger scale, the Pueblo community itself evolves as a social environment. New policy issues arise as the population grows and changes. Some of the issues that have emerged in Pueblo have included the following: (1) guest policies (Should anyone be able to visit for a short time? Should guests be anonymous?), (2) permanent immigration policies (What should be the conditions for the entry of permanent new participants? What kind of orientation should be available to newcomers? Who is responsible for them?), (3) decision-making policies (Who decides when someone's building allowance can be increased? Are there quality guidelines? Who sets them?), and (4) communication forums (Who can or should participate in different discussions?). Each of these questions was prompted by some particular Pueblo situation that unfolded over time. Pueblo, like other online communities, has adapted to its own local opportunities and pressures.

We have emphasized that this particular information ecology has been steered by educational values and goals. But there is no map or book that shows how to adapt a malleable, extensible, social virtual world for a school setting. It is done through ongoing experimentation and reflection. It has been challenging for teachers and school administrators to apply their own practices and intuitions to the design, development, and customization of a novel technology. The process requires persistence and patience.

Here are some excerpts from teachers' reflections on how they want to see Pueblo used in their classrooms:[10]

> "The optimum strategy would be for me to use Pueblo as I would any other resource or strategy I employ in the classroom and be able to adapt it to any part of my curriculum."

> "I see Pueblo as a tool, or more precisely, a set of tools. Success in my classroom would be that the students learn the system well enough to pick and choose from the tools to serve their purposes."

> "I think freedom and choices are important to the success of Pueblo at the classroom level. To that end, I would like my students to work on projects that are important to them. Then, working with me and/or the mentors, they can tie their projects into the curriculum. I would like to see some great things by the end of the year."

> "All students—gifted, average, special ed.—all feel comfortable, confident, and a useful part of Pueblo. . . . All students realize they have something to offer in the Pueblo community, no matter what their ability level is."

Since computer technology (especially global network access) is still new to K-12 education, schools have a chance to think ecologically from the beginning. This is what Longview has done in extending classroom practices into Pueblo. Longview's experience has shown that ecological adoption of technology into the school has several important facets.

Teachers and administrators are involved in the development and implementation of the technology. They make decisions about how hardware will be distributed throughout the school. They help develop training programs for new groups of teachers and students who become involved in Pueblo, so they can pass on their tips and strategies. They consider how the social values of the school setting can be extended and adapted into the online world, and they coach students in their online rights and responsibilities.

They experiment with different ways of incorporating the technology into the classroom as a learning tool, rather than letting the technology dictate to them how it will be used. And they have learned that because there are individual differences among teachers' styles and approaches, there will also be differences in how different teachers adapt and use the technology over time.

Jo Talazus, Longview's principal, reflects on the challenge of sustaining her school's commitment to Pueblo:[11]

> Providing a Pueblo environment is a struggle for an administrator and one which is only worth the time and effort if it highly impacts learning.
>
> Encouraging and supporting teachers as they change the structure of already crowded days is a constant because the pressure to *cover the curriculum* never fades. . . .
>
> Providing enough time for teachers to dialogue as social mores are built is a struggle. What should we allow; what should we not? . . . What are the consequences for misbehaviors? When do the rights of one interfere with the responsibilities of another within the community? . . .
>
> Convincing school boards and administrators that children are not just *having fun* but profiting from that fun can be a struggle in this new world of chat rooms all over the Internet.
>
> Providing technological support to keep donated computers running and the network functioning is a never ending struggle. Finding funds for what is a building initiative is a pressure that never fades.
>
> Keeping a positive climate in a school where some teachers have ten computers and others have none is a struggle. Spreading out the machines may seem equitable but it gives no one adequate access so we've decided to build the system correctly classroom by classroom.

Finding opportunities to train whole classrooms of students is a struggle since thirty computers in a lab is essential; trainers are essential; numerous mentors are essential. Multiplying that by 37 can be overwhelming.

I no longer question if all the struggles are worth it. They are. I know we will one day have built an adequate supply of hardware, have all teachers and students trained, show enormous gains in learning and replicate our success in other schools. In the meantime we forge ahead questioning ourselves at every step, modifying, changing, and growing with each day.

There can be no more eloquent description of the hard work that goes into developing a school-centered information ecology than this. It is complicated and challenging all along the way, but the Pueblo experience is a living example of what can be accomplished when people aim for high-quality, ecologically healthy uses of technology in schools.

9

Cultivating Gardeners: The Importance of Homegrown Expertise

Spreadsheet packages and computer-aided design (CAD) tools are among the most successful software programs. Most people are familiar with spreadsheet software, which is used for budgeting and other financial computations. Engineers and architects use CAD software to create complex drawings of all kinds, from silicon chips to automobiles to building plans. These software tools are widely used all over the world, with installations numbering in the millions. Spreadsheets have been around for about twenty years and CAD systems for a little over a decade. As such, they provide a kind of natural experiment in how software usage in office environments can evolve over time.

The gear and tackle and trim of many trades have been around a lot longer than these software systems—in some instances for hundreds of years. Against this timeline, it may seem early to look at evolutionary processes in spreadsheet and CAD use. But in information technology timelines, ten and twenty years are very long time periods. Spreadsheets are firmly entrenched as essential tools for accountants and others who work in finance. CAD tools are part of the common tool set for mechanical and electrical engineers and for architects.

In a study of spreadsheet users and a separate study of CAD users, Bonnie spent a great deal of time observing and talking to engineers, accountants, drafters, architects, software programmers, small-business owners, and managers to find out how they used software in their offices.[1] In the well-established information ecologies Bonnie studied, relationships among the participants and their technological tools and practices had had time to grow. A fascinating pattern was visible: the presence of "gardeners."

The best way we can think of to start a discussion of gardeners is to quote the words of a "head gardener," as he describes his role in his organization:

> There's an analogy that I think is very good. You know, [here at our company] we have these wonderful hardware systems . . . and we have this wonderful software that we're all using. But we're not really using it to its full potential . . . I'd say that even though we're doing all this great stuff, we're using maybe about 50, 60, 70 percent of the capacity that these tools have to offer. And it's kind of like having this race car in the Indy 500. You know, your boss is giving you this wonderful, fast, fast car, but . . . we don't have a pit crew to help us gas up the car, and fine tune it, and get it to run really fast and beat everybody else. So it's really kind of the development engineers out there racing this car, trying to win this race; getting out of the car, gassing it up, changing the tires themselves, and trying to understand all about this car. We don't have a team in place to help people do this. We just have a bunch of individuals. And I'm trying to build these teams so that we can help each other out more.

This gardener points to the need for people to play diverse roles around technological tools—to cooperate to get the "full potential" of the tools. You've got the car, now where is the pit crew? The gardener draws attention to the time and effort it takes to identify needed roles and see that they are filled. In his organization, it was no simple task to organize a capable bunch of individuals into teams who could help each other out.

Who are gardeners? They are people who like to tinker with computers. They learn the software a little better than everyone else around the office, they're often good at configuring hardware, and they troubleshoot and solve problems when others are stumped. Gardeners like to help other people with technical tasks, as well as learn about computational things on their own.

The term *gardeners* came from a mechanical engineering team in one large company where we studied CAD use. The gardeners there were so designated because they were seen as "growing the productivity" of the company. If you work in an office environment, you can probably think of a gardener in your midst—someone who is a little more adept with whatever software you use and enjoys answering questions.

Gardeners are people who can translate concepts and mechanisms back and forth between the domain of the work and the technology itself. They occupy a special niche in information ecologies, because they bridge the specifics of the domain, with its unique problems and challenges, and the capabilities of the tools used in the domain.

Gardeners are not outside consultants. They are not professional programmers. They are insiders—active professionals in finance or engineering or design or whatever the domain might be. Gardeners know the work, and they know their fellow workers and their problems and frustrations. Gardeners work right alongside everyone else, performing many duties in addition to gardening. This gives them the ability to respond to local needs with sensitivity and understanding.

The head gardener we quoted was looking between the spaces, in the Zen way we talked about earlier, finding the places where the "pit crew" needed to be. It is all too easy to focus on the race and the car, missing the less visible but crucial contributions that surround the central object of activity.

In general, gardeners take on the responsibility of *customizing* software tools for local conditions and *assisting* their co-workers in using the tools. In this chapter, we will explore the contributions of gardeners, highlighting the importance of diversity and locality in the information ecologies they work in. We will draw examples from Bonnie's studies of people who use spreadsheets and computer-aided design tools.

The spreadsheet study involved about a dozen people, all financial professionals of one kind or another who used spreadsheets regularly in their work. There were a variety of work situations represented, including a small, start-up company with eight employees, a very large corporation, a medium-sized manufacturing firm, and independent consultant work. Most of the people interviewed had three to five years of experience with spreadsheets.

The computer-aided design study involved two dozen people from seven companies, which ranged from a three-person architecture design group to a Fortune 100 firm. The people included architects, mechanical engineers, electrical engineers, and industrial designers. Most had been using CAD for at least five years. Unlike the people who participated in the spreadsheet study, many of the CAD users had technical backgrounds and some had taken programming classes, but they were not professional programmers.

As we look at what gardeners are like and what they do in different settings, we may see more clearly how to foster useful links and dependencies in other information ecologies that are in an earlier stage of growth, where people are encountering basic changes in the tools of their trade for the first time or struggling to figure out how to use their tools more efficaciously.

DIVERSITY OF EXPERTISE AND INTERESTS

Gardeners work intensively both with tools and people. In the spreadsheet world, gardeners create "macros" (small programs written in a special purpose language provided by the spreadsheet package), develop sophisticated graphs and charts for presentations, create custom formats (such as a new way to show a value in a spreadsheet cell), and write difficult formulas using advanced functions such as date-time operations. Gardeners may help co-workers "debug" their spreadsheet models—that is, find flaws in the model that are leading to incorrect results. So it is common to find that any given spreadsheet model has been constructed by more than one person. People with different forms of expertise collaborate to develop models. A diversity of expertise in the information ecology is leveraged to get more of "the capacity these tools have to offer," as the head gardener put it.

Gardeners do all of the above, and sometimes they also train interested co-workers in doing these things themselves. Jennifer, an accountant in a well-known telecommunications company, often makes graphs for the controller in her company. She explains her role in the following exchange:

> *Jennifer:* I help the controller. When he's got something on his spreadsheet and he wants me to graph it, I will say, "Can I borrow your disk?" and I will just copy that.
> *Interviewer:* Why doesn't he graph it himself?
> *Jennifer:* He may not have time . . . We are all teaching each other, too. He usually comes to me, "Now how do I do this? I know you did this [so show me]."

There is more than a simple division of labor here. Jennifer helps out with the graphs, but she is also a teacher to the controller. He has observed her spreadsheet models closely enough that he can see interesting things he would like to learn. The interaction of these co-workers is a seamless, flexible way for people to share tasks and knowledge, a hallmark of a smoothly functioning information ecology.

In the CAD world, gardeners also write macros (in the CAD macro languages), "shell scripts" (using a scripting language provided with the operating system), and sometimes programs in conventional programming languages. They may gather these various programs and distribute them more widely at their company, if that is appropriate. Gardeners help set standards so that everyone in the work group uses the same conventions for terminology, documentation, and so forth. Gardeners sometimes evaluate new tools on behalf of their work groups. They answer co-workers' questions, train others who want to learn to write macros and scripts themselves, and generally keep their fingers on the technical pulse of the group.

During all of this, gardeners themselves are immersed in the day-to-day activities of their local group—they are intimately involved in the quotidian pursuits, as well as trials and tribulations, of their local ecology. Their advice, programming, and support is embedded in the natural flow of work in their group, providing a close match between what their co-workers need and what they can offer.

The key to understanding gardening is to notice the diversity of expertise and interests in each work setting. People who are working together on a complex group activity bring their own distinct skills and knowledge. Some are interested in learning more about the advanced functions of the technologies they use, and others simply want the tools to do the right thing without fuss. Those who are good at their own jobs and also have some interest in tinkering with the tools may grow into gardeners.

Here's how a mechanical engineering production drafter describes the gardener she works with. James, the gardener she refers to, is also a mechanical engineering drafter:

> *Carol:* And it does seem like probably just because of different personalities, we all sort of have our area of expertise that when a

certain job comes in, "Oh Harry would be good for this" because technical illustration is really his high point . . .

Interviewer: What's yours?

Carol: Ummmm, I think I'm very good at drafting skills; I know what those views should be, where they should be placed, and I think I'm very good at being a checker on all the specs on the drawing—that when this goes into production, we won't have to scrap parts, we won't have to bring it back for revision after we've gone through with the fine-toothed comb. And that's what I enjoy about it . . . I tackle it like a puzzle, I want to comb everything out and cover every aspect of it and then know that . . . everything's perfect. I like that part. And like I said, James really likes the computer side of it and, "What can I get this thing to do?"

For Carol, the "content" of the work is the main focus of enjoyment, rather than the computer side of it. Rick, a gardener at an architecture firm, also points to the difference between "content" work and the kind of tinkering he does:

Rick: That's why I'm here, because I know that when these guys are designing, they just want to design. They don't want to have to look at the manual, they don't want to have to . . . get into any of that.

The practice of gardening leverages the natural diversity in interest and background knowledge that already exists in local work groups. People often have complementary expertise. Gardening interactions bring people together to solve problems more effectively than if each tried to do it alone or if they hired outsiders who do not understand local conditions. Gardening is an adaptation that is high on accountability. The gardener knows he or she will be there tomorrow, and the next day, and for a good long time, helping solve problems in a context where there is some longevity to relationships.

Here is an example of how different forms of expertise can mesh to lead to a successful resolution of a technical problem. Two engineers in a mechanical engineering consulting firm needed to use a certain mathematical function, a parabola, for a design they were working on. The CAD tool didn't have a built-in parabola function, but the tool was extensible, so this and other functions could be created as a macro by someone with enough programming expertise.

The engineers called on Mark, their local gardener, to help them out. Mark had started as a mechanical engineering drafter and had gradually become more adept at using and customizing the CAD tools. Now his gardening practice was a recognized and valued part of his job:

> *Mark:* Two women were dealing with a lens for a lamp and they wanted to . . . be able to define a parabola easily, and they were doing it somewhat laboriously. As it turned out, the ellipse command that is the standard . . . is a very complex macro that has you defining the major axis, the minor axis, and then it uses the spline command and repeats with all of the points. It's sort of a left-brain, right-brain thing. I can't explain it. I understand when I'm writing it. So I was able to take that and simplify it greatly and make a parabola macro out of it. . . . So by just looking in a standard engineering book, we were able to take the match and I knew the [macro] language and they knew the math, and they just told me how we were supposed to manipulate the numbers that we got out of it.

In this example, Mark and the engineers separately held relevant knowledge—Mark thoroughly understood the CAD tool, and the engineers understood the math. They brought their knowledge together to customize the tool so the engineers could finish designing their lens. Though Mark was not an engineer himself, he was not just a random tool expert, either. As a professional engineering drafter, he knew a great deal about the local engineering practice—the kinds of problems the engineering group addressed and the general techniques and tools they used.

Gardeners have a unique set of skills, because they can speak more than one language—the language of the work domain (such as engineering or accounting) and the language of the technologies and tools. James is another drafter who became a gardener, and he describes this special ability to bridge different domains:

> *James:* The gardener has to have a good working knowledge of Unix . . . because your IT [Information Technology] group is only going to know how to do the administration-type part. They're not going to know the ME30 [CAD tool] part. Your end users [the mechanical engineers] are only going to know the ME30 part and they're not going to know the systems part. So a gardener is like a cross between both worlds. And he's got to be able to communicate what IT is trying to do with the standardized configurations and hardware ordering, and he's also got to be able to speak the

language of the end user who's sitting there saying, "I'm unproductive and I need to be productive real fast." . . . So, it's a juggling act. . . .

Gardeners are co-workers who have "been there before" themselves. They contribute not only technically, but by providing emotional support, as James reveals in his awareness of his co-workers' anxiety about being "unproductive." As we saw with reference librarians, in a healthy information ecology, there is a helping hand at the ready, as well as technical expertise.

This kind of support is not a frill—CAD users work in a very challenging environment. Design and manufacturing processes are changing, deadlines are getting shorter as managers scramble to reduce time to market, and the software tools themselves are sophisticated and constantly evolving. Gardeners dig in and help where help is needed.

The practice of gardening is itself diverse. There is not just one way to be a gardener. For example, some gardeners know about many different tools, while others are expert in only one or two. In an environment where many different complex software packages are used, it is not uncommon to find that there is one person to consult about spreadsheet models, another about making slide presentations, and yet another about putting together a Web page. One of our own gardening friends seems to become the local expert in making online bibliographies wherever he happens to work, since this is something he thoroughly enjoys.

Another area of difference is that some gardeners draw a line about how technical they want to become, and others choose to develop very advanced technical skills through programming courses they take on their own initiative and time.

For example, Ray manages a large staff in a corporate finance department. He has some programming experience, and he is a gardener for his staff. One of Ray's contributions as a gardener was to prepare "targeting templates" for the staff as a way of standardizing the process of setting targets for expenses. The targeting templates are spreadsheet models that have a certain set of prespecified formulas and data types. The templates actually give structure to the targeting activity, by eliciting certain information and organizing it in particular ways. Setting up a template was a spreadsheet customization that leveraged Ray's expertise in the financial processes he wanted his group to follow.

Ray has enough expertise with spreadsheets to set up a good template, but he is not very interested in the programming side of spreadsheet development. He did take an advanced class on spreadsheets, where students learned how to program macros. But when custom macros were needed, Ray got one of the programmers down the hall to do it. He describes this division of labor in the following exchange:

> *Interviewer:* . . . [these menus] look like they'd be pretty useful. And who developed those for you?
> *Ray:* A programmer down in customer support.
> *Interviewer:* Okay, not somebody in your group. You just sent out the work, and . . .
> *Ray:* Yeah, well, essentially, you know, I came at it conceptually, this is what I'd like to see, and they developed it. So [the programmer] made [the menus] interactive, set up the customized use.

The task of making macros is not necessarily beyond Ray's abilities, but it is certainly beyond his interest level, so he enlists someone who is more interested in that kind of thing. He continues to provide creative guidance at the "conceptual" level, which is what he enjoys more himself.

Even in this situation, where the local gardener decided to hand off a task to someone outside his immediate group, there is a strong sense of locality in the gardening practice. Ray did not go outside his company to hire a consultant; instead, he called in a programmer from "down in customer support." This means that he brought in someone who was familiar with his organization and its tools and practices. By relying on someone fairly close by, Ray also increased the likelihood that the special-purpose menus and macros created by the customer support programmer might make their way into spreadsheets used in other local groups. Sharing customizations is one way to share expertise across groups, enriching the practice of both.

Different levels of interest in technology is one area of diversity for gardeners. Another area of difference is that some gardeners play a completely informal support role in their information ecologies, as a sideline to their official job responsibilities, while others are formally recognized for their gardening. Now that we have seen what gardeners do, we will explore some of the ways gardeners can be cultivated.

CULTIVATING GARDENERS

In the spreadsheet settings we studied, gardening was informal, something done "on the side" in addition to other duties. In some of the CAD settings we studied, the informal gardening role had evolved into a formal or semiformal role. Gardeners began to get recognition and release time for their gardening chores. Usually they continued with their regular duties as well, but they had fewer of their old responsibilities as they mixed gardening into their workload. These gardeners appreciated the recognition for their valuable work, which was now visible to management. This recognition was especially important at performance review time.

We are not sure exactly why gardening became formalized in organizations using CAD tools but not those using spreadsheets. It may have something to do with the large investment in CAD software and the high-end workstations that run it, both of which can be very expensive. CAD gardeners often acted as mediators between systems administrators and CAD users, so dealing with hardware configuration, adding new peripherals (such as big plotters and high-end printers), and ordering new, costly software, may have given them a higher profile than spreadsheet gardeners.

Whatever the reasons, we think the model of gardening as a part-time, formally recognized activity is something that others should look at. This is a win-win-win situation—for workers, gardeners, and their managers. Workers like having someone knowledgeable and familiar at their fingertips. Gardeners like having official leave to enjoy their tinkering. Managers, once they see the way gardeners can "grow productivity," appreciate having trusted experts who will cover the many bases of getting complex software to work well.

A gardener can be positioned a little between managers and workers, trusted by both because there is some alignment of goals with both. The gardener is a fellow domain expert, so he or she understands the work problems and is sympathetic to the needs of fellow workers. The gardener can often anticipate concerns and be proactive in handling them, which would not be possible for an outside consultant or support person. The gardener also sees the benefits of standardization, something valued by management, and will try to support these efforts in ways that make sense for the group. Gardeners in more formal roles are responsible for writing and disseminating standard programs and macros at the department, division, or corporate level. The gardener usually takes the extra steps of testing the program carefully, fixing bugs if necessary, perhaps extending the program to make it useful for a wider set of tasks. Gardeners may also research and provide new tools for people in their areas, keeping an eye on the marketplace for new technology developments.

It is perfectly possible for formal and informal gardening practices to flourish side by side. There were several examples of this in the settings Bonnie studied. In this situation, the official gardeners take on the standardization and system maintenance tasks, and the others tend to provide more one-on-one support to their peers.

Though many gardeners pursue their interests without official backing, organizations that want to cultivate gardeners need to think about how to support their efforts. Management can help by paying for books and courses, as well as attending to larger issues such as shifting job responsibilities to acknowledge the time gardeners spend in their consulting practice.

While we studied some organizations with full-time gardeners, the part-time model is attractive because it means the gardener remains closely bound to the ins and outs of daily life in the local ecology. If gardeners are focused exclusively on their gardening activities, they are taken away from day-to-day work in the profession in which they started. A major strength of gardeners is their familiarity with the everyday tasks of the work domain. Over time, full-time gardeners may lose touch with this crucial core of expertise and familiarity. Of course, full-time gardeners may make sense in large organizations that support part-time and informal gardeners as well.

In thinking about how to work gardening into the mix of activities in an information ecology, it is important to recognize that it does no good to create a job description for a gardener unless someone with the right mix of skills and interests is there to fill it. Gardeners have a very special profile. They have domain expertise in the local work activity. They have a strong interest in tinkering with technological tools within the context of their work. And they have good social skills and a desire to help others.

This last requirement is not trivial. A person who is skilled in the work practice and the technology but who is not interested in intensive interactions with others won't have much of a green thumb. We encountered one reluctant gardener who had good technical and work abilities but lacked that essential ingredient of helpfulness to others. Steve was an electrical engineer who worked on a project that involved creating complex simulations. Some of the other teams within the company were doing similar simulations, and they often asked Steve and his partner for help. But Steve's ambivalence about acting in a support role came through in this half-humorous exchange:

> *Interviewer:* So you guys are sort of the experts?
> *Steve:* We're supposed to be, but . . .
> *Interviewer:* (*laughing*) But you're not really? Well, do people come and ask you for help?
> *Steve:* (*laughing*) Well, the thing is, I'm no smarter than any of those guys and . . . we try not to answer their questions.
> *Interviewer:* So you don't like to share what you've done?
> *Steve:* No, no, we will help. It depends on the time . . .

Steve knew that he could help, but he wasn't drawn to sacrifice his own time to share his expertise. It was just not something he was naturally attracted to. Willing gardeners are people who find the activity of helping others rewarding in itself.

Likewise, it is important for gardeners to have a natural affinity for technical things. Rick commented in an interview:

> *Rick:* I don't read manuals, I just start a program and say, "Ah, let's see what it does! Oh, I have this other package that does the same thing." Then you look and see how it does it, and then you get stumped, and then you go look in the manual for reference. And you know, in an afternoon you can figure out everything about a package. That's usually what I do.

Gardeners play a valuable role in their ecologies, emphasizing the importance of diverse contributions and local connections among people and their tools and practices. Gardeners seem to be an evolutionary outcome of the intensive use of software systems over time, as we have seen in many information ecologies we have observed and inhabited. We think it is worthwhile for managers to be on the lookout for potential gardeners, for those who spontaneously gravitate to the tinkering and helping we have described. Cultivating gardeners is one of those looking-in-the-spaces activities that we hope will lead to more diverse, robust information ecologies.

10

Digital Photography at Lincoln High School

Some of the special effects in films such as *Jurassic Park, Forrest Gump,* and *Waterworld* were created with a software program called Adobe Photoshop.[1] At Lincoln High School in San Jose, California, students in Patricia Lynch's Digital Photography class used the same program to create beautiful artwork. In this chapter we visit the information ecology Ms. Lynch created, with emphasis on its grounding in the value of access equity to technology for all students. We describe how the diverse human and technical resources in the ecology helped make it a success.[2]

In the summer of 1995, Bonnie sat in on the Digital Photography class almost every day, from the end of June through early August. She observed and interviewed the students, Ms. Lynch, and the technology coordinator, Cliff Herlth, to learn about multimedia in the high school curriculum. The Digital Photography class is part of an Electronic Arts program created by Mr. Herlth and other teachers at the school. Bonnie also spent time in the other classes in the Electronic Arts program in the spring of 1995.

Lincoln has perhaps the most advanced multimedia program of any high school in the United States. Over a million dollars' worth of high-tech equipment outfit the school, including high-end printers not seen in most offices and video editing equipment better than the local TV station's. Funding for the equipment came from federal and state grants written by teachers over a five-year period under the leadership of Mr. Herlth.[3]

Lincoln is a performing arts magnet school. As a court-ordered desegregated school it is extremely diverse, with students from widely varying ethnic and socioeconomic backgrounds. San Jose itself is one of the most ethnically diverse cities in the country. The summer school program was open to students from all over San Jose, and students from several schools were represented. During the summer school course, Bonnie observed and talked to students from economically privileged backgrounds and students of working class origin whose parents included factory workers, janitors, termite inspectors, security guards, auto mechanics, hair cutters, tire shop operators, and used car dealers. Latino, African American, European American, Vietnamese American, Chinese American, and Azorean American students took the class. Some were immigrants, and some were children and grandchildren of immigrants. A little over half of the students were girls.

Students learned to use an Apple QuickTake digital camera and to alter their photographs in Adobe Photoshop. A digital camera lets the user transfer photos directly into a computer, with no film or developing. Photoshop supports alteration of photos in myriad ways, from very simple to extremely sophisticated, including shading, grading, twirling, warping, collaging, changing textures, colors, sizes, patterns and lighting, and adding text to pictures. Students were able to see their work immediately and get instant feedback from the teacher and other students.

The students' favorite feature of the software was the ability to alter their photographs. In the interviews, the students frequently mentioned the freedom and creativity they experienced in being able to change their photos. Ms. Lynch taught the students to regard digital alteration as an art form—the photographs were not intended to support any kind of "realism." No one was aiming for accuracy or verisimilitude. The original photograph was simply a raw material like paint or stone.

The Digital Photo class was an information ecology grounded in the value of equitable access to technology. Ms. Lynch and Mr. Herlth made sure that everyone succeeded with computers. This was not a pedantic "technology" class that featured word processing or a programming language designed for children. It was an art class in which powerful computational tools, the same as those used by professionals, were deployed to create artwork.

We believe the class was successful for three reasons. First, the artwork, not the computers, drove day-to-day activity. Many of the students were drawn to the class because it was a photography class. They ended up learning to use technology at the same time. Second, Ms. Lynch provided structured lesson plans so no one fell behind. Often technology classes are taught in an "exploratory" mode that leaves many students confused and frustrated. Ms. Lynch ensured that everyone gained competence in using the technology, so they would be able to do their artwork. Third, Mr. Herlth gave Ms. Lynch plenty of technical support so that she could focus on photography, not technical troubleshooting.

THE STUDENTS

In any class, one of the most important ingredients is the students. To get a feel for the class, consider some of the work done by six students. One major assignment was to create a "CD cover." The students had to design the front, back, and spine of the cover for a musical group they knew of or invented. They had to use at least one photograph they had taken (as opposed to using scanned images).

Amber Doyle's CD cover was dedicated to a local group called the Cranes, whom she admired and often went to listen to. She photographed a park bench in the Rose Garden, a city park near Lincoln High, to get a striking fleur-de-lis pattern for the cover. Amber colored the cover in lovely, watery shades of blue and green. Her work in Photoshop involved color manipulations, reshaping some parts of the image, and adding text. Amber was working hard in school, as she hoped to get a scholarship to go to college. She was considering joining the Navy, a family tradition. Amber did not have a computer at home and learned everything she knew about computers in school.

Francisco (Cisco) Martinez created a CD cover showing a photo of his good friend Daniel on a skateboard and Daniel's little brother on a collaged-in skateboard. The manipulation involved adding a skateboard for Daniel's brother and smoothing the edges to make the images blend, which took careful work. Cisco thoughtfully chose the colors for his CD cover, trying a number of different color combinations. His cover was devoted to "house" (a kind of music, like techno but faster). Cisco enjoyed math and planned to be an engineer. He had a computer at home that he used to play games and make simple drawings, such as for party invitations. Photoshop was a new and fascinating tool for him.

Liz Victoreen was an excellent student, taking a heavy load of advanced placement classes. She was an experienced computer user with her own computer at home. Liz's CD cover was puckish and fun, a postmodern look back at the 1960s. It showed an old Volkswagen Beetle rendered in psychedelic colors, and it was titled "The Grateful Dead," although the song list contained Beatles' songs. (This was deliberate; Liz and her generation are very familiar with the music of the previous generation.)

Holman Vilchez did not have a computer at home, but he became proficient at school. He hoped to be an architect. Holman's beautiful CD cover was based on an oil painting he did from a picture he found in a book. It showed a rural church on a dusty road, perhaps in Mexico. He took a 35 mm photo of the painting and scanned it in. In the upper corner of the digital photo, Holman collaged in four separate photographs of his family members and himself, including a picture of his sister at her *quinceañera*, or fifteenth birthday party. The CD cover was called "Mi Pueblo."

Gilbert Fregoso hoped to be an engineer. His imaginative CD cover started with a photo of a volcano that he found in a book and liked very much. For the photo he himself took, Gilbert shot a fire extinguisher hanging on the wall in the classroom. He took an ordinary object and altered it in Photoshop. It looked like a fire extinguisher designed by Ralph Lauren by the time Gilbert was done. He used the advanced layering techniques he learned from watching the Photoshop instructional video Ms. Lynch showed during class. Gilbert combined the volcano and the fire extinguisher artfully in his cover, which he called "Hot Hits of the 90's." His artwork was typical of that of many students who took the rather sterile surroundings of the school and made them interesting in their digital photos.

Gigi DaCosta lived in the Azores as a child and spoke fluent Portuguese. She studied hard and worked after school at a men's clothing store. Her CD cover, called "Ridiculous Dreams," was a picture she shot of Cisco Martinez (who sat near her in the class). Cisco was lying on a bench in the courtyard outside the classroom. The picture was beautifully composed, with a sophisticated sense of light and shadow. Gigi altered the photo so that Cisco appeared to almost float over the bench, as in a dream, surrounded by shadows.

SCHOOL-CENTERED VALUES: ACCESS EQUITY

The Digital Photography class, unlike many classes that intensively employ high technology, was not the preserve of a small group of technically minded middle-class boys. It was full of girls, students from all social classes, and members of many ethnic groups. Many of the Digital Photography students did not have computers at home. A few had never used a computer before they took the course. None had used a software program as sophisticated as Adobe Photoshop.

While Ms. Lynch provided the leadership for promoting the value of access equity in the information ecology, the students also believed in this value. They recognized that experience with computers would help them in their futures.

Many thoughtful people, such as Larry Cuban, a noted educator, believe that computers are not appropriate in the classroom.[4] We think such commentators have a point when computers are not designed into the curriculum properly, as is often the case. But there is a larger point being missed in their argument. Computers *are* important in today's world. Occupations such as physician, nurse, engineer, scientist, architect, accountant, and small-business owner, to name a few, require computer skills. Factory workers use computers, and so do insurance agents and transit workers and secretaries. Students who do not have computers at home are at a disadvantage if they do not work with them at school. We agree with the students in the Digital Photo class that having confidence and skill with software tools is important preparation for many kinds of work. These skills will come when classes such as the Digital Photo class incorporate computers as tools—when computers are used ecologically, to attain some valued goal within the ecology.

"THE JOBS OF TOMORROW AND STUFF LIKE THAT"

While adults argue about whether schools should have computers, the students in the Digital Photography class were unequivocal on the matter. Virtually everyone interviewed said that computers are important and that they should be available in school.

For example, Amber worried about getting computer skills not only for herself, but also for her brother:

Interviewer: Do you have a computer at home?

Amber: No.

Interviewer: So everything you've learned about computers is being in public schools through the years?

Amber: Yeah.

Interviewer: Do you feel that's an advantage for you?

Amber: Kind of but, like, I was really disappointed because in middle school I didn't learn anything about computers. They really didn't include that in the curriculum at all.

Interviewer: Really?

Amber: Not at all. . . . I mean, in sixth grade in math, they would take us there and let us play on the computers a little, MathScape. For like English I think that should be included and it wasn't. It wasn't here [at Lincoln] but, there's classes that you can take here. And there you couldn't take those classes unless you were already advanced.

Interviewer: So you think they should have had more than they do?

Amber: Yeah, I really do. But, hopefully they're changing, like changing it a little bit so that there's more opportunities for the people who to go to Hoover [the middle school] because my brother goes to Hoover and I would like him to be in that because he's interested in computers right now. And like graphics and stuff, he's really interested in that and I, I think that's good, but I was hoping that they offer that now for him, because they didn't offer that for me.

Interviewer: Now are you taking any of the other computer classes here, like the Video or Multimedia or any of those?

Amber: No, not yet. Because right now I'm not really worried about my electives and stuff. Basically I'm paying a lot of attention to things I need to get done and then hopefully my senior year I will have most of the things I need to be done, done so I can work on my electives and like computer skills.

Amber had a very sensible attitude toward her school career, balancing different objectives and organizing her schedule so that in her senior year she would have time to acquire some experience with computers.

Another student, Ryan Hoskin, had the following to say:

Interviewer: And do you like photography? Is this capturing your interest?

Ryan: Actually I didn't choose this class. This was given to me because there was no more room anywhere else, and I wasn't really thrilled about it when I started, but I really like this class now. So I'm signed up for a couple more classes involving this next year.

Interviewer: Okay, that's good. What do you like about it?

Ryan: Just that, like, it's easy and it's fun and it's like this is the kind of stuff that's going to prepare you for the jobs of tomorrow and stuff like that.

Interviewer: You mean photography or computers?

Ryan: Computers.

A total of sixteen hundred students in grades 9–12 attended Lincoln High School at the time of the research. In the spring of 1995, every prospective sophomore, junior, and senior signed up for a class in the Electronic Arts program for the 1995–96 school year—many more than could be handled. This is a strong bottom-line indication of student interest in this program of elective classes. Lincoln is a school that cannot be described as privileged, wealthy, or oriented toward math and science, yet there was intense interest in courses that gave students access to advanced technology.

Not all students gravitate to exploring computers on their own, and not all students take advanced math and science classes in high school, where they might encounter the power of computers. If computers are not experienced through the arts or some other non-math-science area with intrinsic interest of its own, a smaller group of students gets the benefit of learning how to use computational tools.

It is interesting to look back at the comments of Gigi, Amber, and Joe regarding their evolving understanding of the relationship between their interest in photography and the computer tools. All were surprised at the power and utility of the tools as they began to experience them. As Amber said, "I didn't think I could do that until, like, I started doing it."

MAKING DIVERSITY WORK

There is a useful comparison to be made between the Digital Photography class and two other classes in the Electronic Arts program at Lincoln, the Multimedia and Digital Video classes. The Multimedia class was about 80 percent boys and had considerably less ethnic diversity than did the Digital Photography class. The Digital Video class—which was fed from the Multimedia class—was nearly all boys. The students in the Digital Video class did incredibly good work (including several public service announcements shown many times on the local San Jose television station). But the class was primarily the enclave of a small number of technically oriented boys.

Why did the Digital Photography class at Lincoln succeed in attaining access equity while the Multimedia and Digital Video classes did not? It was not because anyone wanted to keep girls (or anyone) out of any class. It was because of the way the classes were structured. It takes thought and sometimes experimentation to create a situation in which the application of values comes to fruition.

In the Digital Photo class, three factors came into play. First, it was an *art* class—a photography class—rather than a technology class. Second, Ms. Lynch provided a structured curriculum in which students were not just expected to "get it" without systematic instruction. No student floundered trying to understand how to operate the computer, because Ms. Lynch did not let that happen. Third, Ms. Lynch did her work in a diverse information ecology in which she herself was supported amply in installing and maintaining the equipment in her classroom and in getting help with technical matters whenever she needed it.

Art, Not Computers

Information ecologies support real human activities. They are not about technology for technology's sake. Nearly all of the students who took the Digital Photography class were more interested in photography than computers. Some did not even know that the "Digital" part of the course title had something to do with computers. When they found out, they were still interested, since they really wanted to learn about photography. Some students were pleasantly surprised at how much they enjoyed the digital aspects of the class.

For example, Liz said:

> *Liz:* I was really taking this course because I thought it would be like traditional photography with the dark room and all that. I was disappointed when I found out that it wasn't that, but my friend, she took this class last year. She said it was fun. . . . And also I just

like photography and I didn't have a chance to take it at my school last year because . . . like, I have to do certain electives so I just wanted to try photography.

Interviewer: How do you like the digital?

Liz: I like it. I mean the cameras are easier to—I know like what I'm gonna get, because you know—I mean sometimes I've had really horrible experiences with film. Like once I took a whole roll of film in another country which I can never take again without the film in there. It makes me mad when I don't know what I'm gonna get. But with a digital camera I do, so that's a little easier for me.

Amber explained:

Interviewer: Okay. And so why did you want to take this class?

Amber: Because my friends took it. Candy and Joel they took it, like through high school, and so I wanted to take it because I'm interested in photography and stuff. I wasn't particularly interested in digital photography, but I mean it has something to do with it so I decided to take it.

Interviewer: It had something to do with photography you mean?

Amber: Yeah, so I decided to take it.

Interviewer: And how do you like it?

Amber: I like it. I enjoy it a lot. It's really fun.

Interviewer: Yeah, good. And do you like the digital part even though that isn't what you actually signed up for?

Amber: Oh, yeah, because I can like alter the pictures and stuff. I didn't think I could do that until like I started doing it. It works out nice. I like it.

Gigi expressed similar views:

Interviewer: And so, how about the digital? Did you want to do the computer-based? How did you feel about that?

Gigi: I didn't know that this was going to be digital, but when I found out, I really got excited.

Interviewer: Oh, really? Oh, good, because some of the students thought they were doing darkroom.

Gigi: Yeah, that's what I figured, too, but when I found out, this is like better, I think.

Interviewer: And why do you think it's better?

Gigi: Because you get to change the picture around. You can, like, do almost anything. . . .

Interviewer: How do you like the cameras that you've been using in here?

Gigi: I like them. I thought they were going to be more complicated than that.

Interviewer: Yeah? So do you like the fact that it's easy to use?

Gigi: Yeah. And it's weird. I thought it was weird how you can get a picture into the computer.

Interviewer: You didn't expect that?

Gigi: No. I was like, how do you do that?

Interviewer: And so do you like that, being able to do that?

Gigi: Yeah, like you can get a picture that you didn't like and fix it up to make it the way you want it.

Amber and Gigi both mentioned that they didn't think they could alter pictures. The possibility of doing such a thing with a computer did not occur to them. Gigi thought the cameras would be too "complicated." Many girls echoed these themes. They felt that they did not understand the technical possibilities, and they did not think they themselves could use and control such technology. The possibility of using technology was opened to them because they found themselves in a class in which the technology was embedded in an activity—photography—that they wanted to do.

Not only the girls were surprised at the possibilities of digital photography. Joe Polizzoto, a senior, made similar observations:

Interviewer: And why did you want to take this class?

Joe: I like photography. I had taken a photography class last year [in summer school] and just learned the basics. . . . [This summer] I didn't have the class that I asked for which was Ethnic Studies so this was, this looked like something that maybe I could take. I didn't really know what digital, you know, how to do this, so I was interested, but like I said, I probably would have, I was hoping I could take Ethnic Studies, the other class, but they had to drop it so I just had to pick a class so this was, like, interesting.

Interviewer: So how does this compare with the other class that you took in summer school, the other photography class?

Joe: Well, the other one was just, we used regular cameras and we used a dark room and developed our own film, and this one is like—awesome! You get to use computers and . . . it's just really

different in that you use your creativity and stuff, like, a lot more.
Interviewer: In what ways can you use your creativity more in here?
Joe: Okay. On the computer you can like, you can like take a picture of one thing and then another and then maybe put them together like what you're imagining in your mind, and you can just really, like there's a lot of different like stuff you can do with, like, each picture you can maybe twirl it around or, there's a lot of, you can really experiment, which is really the whole thing to me, which is me. You just experiment and like figure out you think looks good and build on that.

Students who are not particularly attracted to computers for their own sake are able to learn to use advanced technology with ease when it is embedded within an activity that they find meaningful and intrinsically interesting, such as photography. The students quickly became hooked on the creative possibilities of digital photography. Learning to use the technology flowed from their interest in getting to "use your creativity and stuff, like, a lot more."

A content-based focus opens up a class to a much wider range of students than a class geared to learning technology per se. It is especially apt for students to learn the same tools professionals use, not watered-down "educational" versions. Students take pride in learning to use real tools, whether they are digital imaging tools, spreadsheets, desktop publishing tools, scientific simulations, or any other kind of tool used in the real world.

The Curriculum

In addition to focusing the class on the art of photography, Ms. Lynch provided a carefully structured curriculum in which students were prepared for their work at each step. The way Ms. Lynch wove together the lessons on photography and the use of computers can be seen in a brief chronological account of the class.

The class began with an explanation of the principles of photography and examples of excellent work. The students viewed *The Black Stallion*, a film directed by Francis Ford Coppola, to heighten their visual sense. Ms. Lynch asked students to watch how camera angles, backlighting, and framing were used in the film.

Students received a sheet with information on traditional cameras, so they could learn about the cameras' technical features. Then they were given an assignment to take twenty-four shots with a 35 mm camera, centered around several themes: "In My Life" (something that stands out about the student), "Pure Color" (colorful things in bright light), "Patterns" (abstract repetitions), "Family" (anyone in the student's family), and one theme to be chosen by the student.

Ms. Lynch advised her students to "be aware of the full frame of the picture," try different camera angles, avoid taking unconsidered snapshots, think about perspective, and avoid clutter. When shooting people, she asked students to be aware of "where you set them," paying attention to what is behind the person. A key message of the class was to "be aware." The assignments gave the students analytical practice in seeing more perceptively. At this point, Ms. Lynch also talked about the work of Dorothea Lange and Ansel Adams.

The students were assigned to make a camera obscura with black construction paper and glue. A discussion followed. "Why did you build this camera obscura?" asked Ms. Lynch. Nineteenth-century artists used them, she noted. Ms. Lynch explained how light moves, and she described how black-and-white and color photos are processed. Students viewed slides from the Kodak company on the principles of photography.

As computers were introduced into the course, Ms. Lynch set up a social context for their use. Students read and signed a contract that specified some expected behavior toward the computers in the classroom. Students agreed not to open the system folder, change control panel settings, or use profanity in file names. Students were matched with partners, as they had to share computers and cameras. There were twenty-five computers, twelve cameras, and thirty-two students in the class.

The students' first experience with digital photography was having their own pictures taken by the teaching assistants (two students who had taken the class before). The pictures were used for a "Press Pass," an ID that permitted students to walk around campus during class time to take pictures.

As part of the normal routine of an art class, students displayed their work to Ms. Lynch and the rest of the class to be critiqued. The group talked about what to do when people are "camera shy," which led into a general discussion of the ethics of photography. "People have a right to refuse a picture," observed Ms. Lynch.

Ms. Lynch took a group picture with a digital camera. She talked about computer basics, such as floppy disks, files, hard disks, and networks. She introduced Photoshop by editing one of the group photos, putting horns and a mustache on one of the students. This was an old trick, but the students thought it was funny.

After Ms. Lynch gave the students an overview of how to use Photoshop to manipulate photographs, she moved into the details. The students received a handout showing each of the ToolBox icons in Photoshop, so they could learn basic operations.

Now the students were almost a couple weeks into the class. By the time they reached the digital part of the curriculum, they had already thought about the history of photography, great photographers, principles of photography, traditional tools, and the ethics of photography. The computer was introduced when the students had sufficient grounding in the basics of the art form.

Students delved into Photoshop. They tried out shading, grading, twirling, textures, and colors. Some tried the Photoshop tutorial. They learned about cloning, rubber-stamping, filling a pattern, adding text to an image, and many other operations in Photoshop.

If the students needed help, they could put a red cup on top of their computer monitor as a signal, and Ms. Lynch or a teaching assistant would stop by and help them. Sometimes small things like the red cup system make the difference between an information ecology and a roomful of machines. The students had an easy, socially sanctioned way to ask for help. The low-tech red cups were cheap and effective. Most of all, they were a visible way of saying, "It's OK to ask for help. We expect you'll be asking for help." Students used the red cups all the time.

After another week, the students were out shooting with the digital cameras. During this week, students learned that in 1826 it took eight hours to expose a photograph. This was a big advance compared to sitting for a portrait. The students learned about Frederick Evans, Lewis Hine, Alfred Stieglitz, the Dadaists' montage photographs, Dorothea Lange, and Walker Evans.

Brian Reilly, a researcher at Apple Computer at the time, gave the class a demonstration of QuickTime VR (Virtual Reality) so students could see what the future of photography might hold. QuickTime VR is a technology for "stitching" together still photographs to create a video. It has been used, for example, to create the experience of walking around inside a museum. Ms. Lynch skillfully and subtly placed current digital photography techniques between their predecessors and successors, infusing the course with a sense of history.

The students spent the rest of the course working on a variety of assignments, including the CD cover and an "advertising" shoot. They could find a subject for an ad themselves on campus or use props Ms. Lynch kept in the classroom, such as motor oil cans and cosmetic boxes. Some very persuasive motor oil ads emerged, with old Chevys and cool teenage automobile owners. In these later assignments, students also learned some of the advanced features of Photoshop. They received lots of feedback on their photos from Ms. Lynch throughout the course.

In any discipline it is necessary to master technique. What keeps the dancer at the barre or the batter in the cage is the vision of the activity that is enabled by all the drill and practice. Teachers must give students the requisite computer skills to empower them to realize their larger interest in a subject like digital photography. With a valued goal in front of them, students acquire technical skills that might seem boring and low-level if these skills were taught for their own sake. Many "technology" classes founder on the boredom of learning a disconnected skill that will supposedly lead to "computer literacy." Students pay attention and learn when the larger goal is obvious, attainable, and valued.

Nurturing and Support in an Information Ecology

Sometimes Ms. Lynch had to do more than provide an overarching goal for students. She had to help students overcome fear of failing with technology. She took the time and energy to encourage students who might not think of themselves as users of advanced technology. Ms. Lynch related the following story:

> For some of these students, they've never really had any kind of experience on a computer. One story in particular, I have one

student in here who did the traditional [photography] class [which Ms. Lynch used to teach], and I suggested she take this class, knowing I would be teaching. I said, "You know, wouldn't it be neat if you could?" And she says, "You know, I am scared to death of computers." And I said, "You know, I don't know a lot about them either. Come on, let's try it." And she said, "Well, Ms. Lynch, I really don't know." Well now she just loves it!

In addition to carefully coaching the students on the basics of photography and the operation of the computer hardware and software, Ms. Lynch attended to the wider information ecology by emphasizing etiquette and decorum. No profanity in file names was allowed. No messing with the operating system. No gum chewing. Ms. Lynch enforced strict rules to keep the classroom clean, and it was always spotless. She was often seen getting out her carpet sweeper the moment any dirt appeared on the new carpet. It is always important to keep a classroom clean, but computers are finicky about dirt and dust. The students responded by showing respect for the room and helping Ms. Lynch maintain it.

Students could use the red cups as an approved way to seek the attention of Ms. Lynch or the teaching assistants. The cups are a good example of technical diversity in an information ecology. Sometimes the right tool for the job is very humble.

Mr. Herlth, the technology coordinator, was an important resource in the ecology of the Electronic Arts program at Lincoln. He was an industrial arts teacher who had always had a strong interest in computers. Mr. Herlth found it quite natural to move from Industrial Arts (wood, metal, auto, drafting) to digital technology. As he noted in an interview, a computer is just another tool. Mr. Herlth provided crucial technical support to Ms. Lynch, who did not know a great deal about computers when she began the Digital Photo class.

Ms. Lynch has been a teacher for many years. She has a degree in art. She taught traditional photography and yearbook in the past, as well as other subjects earlier in her career. In 1987, she was voted the District Teacher of the Year. Her ascent to doyenne of digital photography was swift—the first time she taught the class was during the 1994–95 school year, when she knew nothing about digital photography and had only moderate experience with computers. In addition to keeping up with more than thirty teenagers in each of her five classes, she managed a lab of twenty-five Macintoshes, twelve digital cameras, and a very high-end printer.

It was the partnership between Ms. Lynch and Mr. Herlth that made the digital photography class an immediate success. Mr. Herlth helped Ms. Lynch maintain and troubleshoot the equipment. He gave her ample informal coaching in the use of the software. As Ms. Lynch observed, "I feel one of the biggest pluses of any kind of program like this is the fact that we have Cliff, who is a resource, to really help you, and he really understands." Mr. Herlth not only had strong technical skills, but he enjoyed sharing his knowledge. He had the teacher's gift of knowing how to present material in just such a way that the learner could grasp it readily.

We believe it is easier to find teachers who can grow technically than to find technical people who will somehow come to understand the pedagogical issues that confront teachers. Teachers understand the needs and habits of young people, the pace of a school day and year, activity flow, and student attention spans. They understand how to manage the zillion pieces of the puzzle that make a classroom work: sign-up sheets, contracts, handouts, groupings of students, assessment, communication with parents, and many other things. Just as we saw gardeners forming crucial links between their co-workers and the capabilities of the tools they were using, teachers are the natural bridge between students and technology.[5]

Ms. Lynch observed that some of the nurturing of the information ecology came from the *students*. Input from students is often overlooked in the design of innovative educational experiences. Ms. Lynch drew on and encouraged diverse sources of input and support in her ecology. She felt she could waive her role as all-knowing authority, and she found the students helpful and responsive. She concluded:

> So it was really learning by doing. I didn't get any formal training in Adobe Photoshop, so I was nervous in a sense, and yet felt that it was really important to expand my knowledge. . . . It was very scary, and yet I presented to the students as if I really didn't have all the answers and I really was learning along with them. And they've been wonderful.

Ms. Lynch had a nice way of describing the rewards of participating in a thriving information ecology: "It stretches your brain beyond anything!"

11

A Dysfunctional Ecology: Privacy Issues at a Teaching Hospital

Up to now, we have concentrated on vibrant, healthy information ecologies. But in some information ecologies, the smooth relationships among people, technologies, practices, and values break down. Things can get out of whack, and it is important to understand some of the root causes of dysfunction. In this chapter, we discuss a breakdown in one particular information ecology as an example of how things can go awry when a new technology is introduced.

In 1992, Bonnie spent several months studying the introduction of an experimental monitoring system in a large, well-known teaching hospital.[1] The new system was used for remote broadcast of surgical events during neurosurgeries. The idea behind the system was to allow a neurophysiologist (a specialist in nervous system physiology) to monitor several operations at once, by seeing and hearing information from each operating room from a computer in his own office or at other locations in the hospital. If a surgery reached a critical stage and needed the neurophysiologist's physical presence, he could reach that operating room quickly and would already know what had been taking place there. Without such a system, each neurophysiologist could only move between the physical locations of a few operations. Neurophysiologists are scarce, expensive resources. It was hoped that making their expertise available to more operations would improve the outcomes for patients undergoing brain surgery.

To learn about the setting in which this new technology was used, Bonnie spent many hours in the operating room, watching the proceedings during several operations. The staff were accustomed to visitors, as they are in most teaching hospitals. Like everyone else, Bonnie donned "scrubs" (soft, loose, cotton clothing) and a mask before entering the operating room. Most of the time she sat on a stool and watched the operation. Because there is so much routine work in a brain surgery, there were many opportunities for quiet conversations with operating room personnel. Bonnie spoke with anesthesiologists, nurse-anesthetists, circulating nurses, neurophysiologists, and neurotechnicians. Only the scrub nurse and attending neurosurgeon could not be disturbed. She conducted a more formal set of interviews with all operating room personnel outside the operating room and was able to catch the surgeons and scrub nurses in that round. It was important to interview everyone, because it is the combined expertise of the very different specialties—neurosurgery, anesthesiology, neurophysiology, and nursing—that make an operation a success.

In this chapter we take a detailed look at how the remote monitoring technology was deployed in the operating room and beyond. We discuss the useful capabilities of the new technology, and then we see how its uncritical, unreflective installation stirred up problems by violating some of the values of the local ecology. Though no one intended for the consequences of this change to be far-reaching or troubling, the monitoring system radically transformed the information ecology of the operating room.

A QUICK TOUR OF A NEUROSURGICAL OPERATION

Neurosurgical operations correct nervous system problems and typically involve surgery on the brain or spine. There are many different people and technologies at work, and members of the surgical team have to coordinate their activities very carefully. To see how this coordination works, let's look at who is involved in a brain surgery and what they do.

In a typical operation, the patient is tranquilized, anesthetized, and connected to a variety of monitors and drips. The attending anesthesiologist has already planned the general course of the anesthesia to be used for the operation, and he or she is usually present during the "prep" period. The anesthesiologist works with the nurse-anesthetist and/or a resident anesthesiologist (a doctor being trained in anesthesiology) to administer the anesthesia and insert the appropriate intravenous lines for blood and a catheter for urine. After the initial setup, the attending anesthesiologist generally leaves the operating room to attend to another operation or take care of other tasks. The resident and/or nurse-anesthetist monitors the patient's basic physiological functions: heart rate, blood gases, blood pressure, breathing, urine concentration. The attending anesthesiologist returns to the operating room when necessary. She can be reached through her pager, and she makes check-in visits to see how things are going.

While the patient is being prepared for surgery, the nurses are busy setting up the operating room: arranging instruments on the sterile table to be placed near the operating table, checking supplies, and putting the patient's X rays up on the wall.

After the prep, the patient is "opened"—that is, the incision is made—by the resident or neurosurgical fellow (doctors at different stages of training in neurosurgery). The resident or fellow then continues to cut and drill until he or she is down to the point in the brain or spine where the most delicate surgery is required. For example, he reaches the point at which some brain tissue must be "picked through" to reach a tumor, aneurysm, or blood vessel compressing a nerve.

When the opening is complete, there is a quick rustle of anticipation and change of atmosphere as the attending neurosurgeon enters the operating room. Now the difficult task of removing a tumor or repairing a nerve begins. The attending neurosurgeon works with a different set of tools, faces less predictable outcomes, and is responsible for the most perilous and difficult parts of a surgery. Often the neurosurgeon can only work while peering through a microscope to see what he is doing. The structures he manipulates are tiny, and the shape of the tissue complex and convoluted.

This "microsurgery," as it is called when the microscope is in use, accounts for only a portion of certain difficult operations. But it involves the trickiest maneuvers a neurosurgeon undertakes. At this time, no one else in the operating room has a clue as to the surgeon's current whereabouts and actions except through his verbal reports—a kind of play-by-play of a scalpel picking through tissue.

Throughout the surgery, the scrub nurse hands the surgeon the instruments and supplies as he requests them. The circulating nurse makes sure that the scrub nurse has everything she needs. When the attending neuro-surgeon finishes his work, the patient is "closed"—that is, the incision is repaired—by the resident or fellow. The patient is revived from the anesthesia in the operating room. He is asked to wiggle his toes and say something. He is then wheeled to recovery.

Long stretches of an operation, especially during the opening and closing of the skull, permit a relaxed informal atmosphere in the operating room. Perhaps the radio is playing, people talk about the latest movie they saw, and they tell a few jokes.

THE EXPERIMENTAL MONITORING SYSTEM

A surgeon's verbal blow-by-blow account of his progress during surgery has its limitations, especially during tense moments when he is concentrating on not inflicting lasting damage to the patient. While a micro-surgery is in progress, a visual view of what the surgeon sees as he works in the brain can provide everyone else in the operating room with a much better sense of the progress of the operation than they could get from the surgeon's verbal reports alone. In some hospitals, including the teaching hospital we studied, a system has been set up to allow other people in the operating room to see what is going on by watching a TV monitor.

To give everyone a common, shared view of the surgical events, a video camera is hooked up to the microscope used by the surgeon. In technical terms, the camera is comounted with the optics of the surgeon's micro-scope. This means that the camera captures a video image of whatever is in the surgeon's field of vision. The live color image is displayed on a monitor hooked up to a cable TV link. Everyone in the operating room can see what the surgeon is up to by looking at the monitor.

This video technology has been in existence (though not universally available) for over twenty years. It is an indispensable part of operating room activity in many hospitals, and hospital staff have grown accus-tomed to its use. With the video, the entire neurosurgical team can see what the neurosurgeon sees. As one scrub nurse in our study said, the video is "the only indication we have of what's going on in the [patient's] head." For example, with a quick glance at the TV monitor, the scrub nurse can be clued in that the surgeon is approaching a time in the operation when a clip will be needed. Or she may see the surgeon nick some tissue, in which case a cautery device will likely be called for to repair the nick. Each person in the operating room interprets the video image according to his or her particular expertise and responsibilities.

The experimental monitoring system in the hospital we studied went well beyond the video capability we have described so far. In addition to the silent image of the operating site on the TV screen in the operating room, the system had an innovative facility for *remote broadcast* of video, audio, and quantitative data from the operating room instrumentation. The audio consisted mostly of the surgeon "talking through" his work during the operation, as we have described, but it also picked up ambient noise and the voices of those closest to the microphone attached to the video camera.

Why would anyone want to broadcast operating room audio and video outside the operating room itself? To answer this question, we have to look at the work of a key player in the operating room: the neurophysiologist. It was neurophysiologists who developed the monitoring facility in the hopes of making their own jobs better. Their goal was to be able to do some of their work in locations *away from the operating room.* What the neurophysiologists dreamed up was nothing less than a radically altered information ecology in which their own activities could be transformed through the technological support provided by a sophisticated new system.

Neurophysiologists monitor the patient's neurophysiological responses during an operation. They feed information about the responses back to the neurosurgeon and anesthesiologist if they suspect a problem. Neurophysiologists help make sure that the patient does not emerge from the surgery blind or deaf or with some other neurophysiological deficit. Neurosurgery today is relatively safe in terms of mortality, but the art of picking through the brain is still an art, and things can go wrong at unexpected moments. Neurophysiological monitoring has been successful in reducing damage to patients by constantly tracking the activity of the central nervous system to see that it stays within acceptable levels. During many operations, there is a high risk of damage due to the anesthesia or surgery itself. For example, it is possible to cut, stretch, or compress a nerve, or to cut off the blood supply to parts of the brain.

Neurophysiological monitoring helps prevent such events. It has made it possible for neurosurgeons to perform more difficult and daring operations. Neurophysiological monitoring is available only in some hospitals, and it is used during operations in which the patient is at high risk.

The essence of neurophysiological monitoring is analyzing data taken from instruments attached to the patient's body. Electrical data showing nerve responses appear as plotted line graphs on a computer screen. Basically, neurophysiologists sit around watching graphs, looking with a highly trained eye for unusual patterns in the data that might spell trouble.

Some neurophysiologists believe that they can use their time more effectively by monitoring several operations at once or working on other projects during the trouble-free parts of operations. They simply do not need to be physically present in the operating room for the lengthy duration of an entire neurosurgery (which typically lasts from five to twenty-four hours). Doing neurophysiological monitoring remotely allows them to serve a greater patient population, spreading their scarce expertise over a greater number of operations.

Computers make this possible. The quantitative information in the plotted line graphs can be broadcast to computers outside the operating room, when the computers are connected over a network. In this way, a neurophysiologist can monitor the graphs from his office, another operating room, or a conference room, as long as he has access to a networked computer in that location.

With this networked technology, neurophysiologists can spend part of the day in the various operating rooms and part of the day in their offices or other locations, monitoring the computer displays. They can monitor as many as six operations at once, though a backup is always assigned to help in case of overload. Neurophysiologists also work with a subordinate called a neurotechnician, who is always present in the operating room and always monitoring the graphs.

When the neurophysiologist is not in the operating room, he communicates with the operating room over a telephone, and he carries a pager so he can be called quickly. If the neurotechnician suspects a problem during the course of the operation, she reports it to the surgeon. She may also telephone the neurophysiologist if he is not in the operating room and has not recently phoned in. The neurophysiologist then returns to the operating room to evaluate the troubling data and talk directly to the surgical team.

The networked system showing the plotted line graphs, located in remote locations away from the operating room, was pretty innovative in and of itself. The expanded experimental system, with its audio and video multimedia capabilities, was pushing considerably further. It added a new—and, as it turned out, problematic—dimension to the remote facilities. The new system fed *all* the data of interest to neurophysiologists outside the operating room, including operating room audio and video, in addition to the instrument data.

The neurophysiologists went to a lot of trouble to create this system. They wanted the audio and video to be broadcast so that when they were working from remote locations, they would have crucial background information that helped them interpret the instrument data in the plotted line graphs. Scanning for disturbing patterns in graphic data is a challenging task. It is not just a matter of reading the numbers and seeing a clear indication of a problem. The neurophysiologist needs to interpret the data in the graphs, in order to make a judgment about what they mean. Difficulty in interpreting neurophysiological data arises from signal noise, the type and amount of anesthesia used, surgical events, and random variation. The audio and video provide an important source of information neurophysiologists can use for making better inferences in a highly interpretive task.

Any extra information about possible problems is useful to the neurophysiologist. For example, when a retractor is placed, a delayed physiological response should not necessarily be attributed to nerve damage but may have been caused by the retractor itself. The neurophysiologist wants to know when the retractor is placed so he will not generate a false alarm. The graphs alone won't tell him this, but the video image will. By seeing the graphs and video side by side, the neurophysiologist has a more complete understanding of the progress of the operation. Over the networked monitoring system, the neurophysiologist can see the video data just as he can when he is physically present in the operating room. Instead of glancing over at the monitor from his position near the computer with the plotted line graphs, the neurophysiologist can look at the video image on the computer sitting on his desk.

The audio adds even more useful information. The microphone is mounted on the microscope, so it picks up the surgeon's voice, the voices of the people closest to the surgeon, and some ambient noise in the operating room. Through this audio channel, the neurophysiologist can hear some of what is being said in the operating room, and he can also get a sense of the overall emotional atmosphere there. The neurosurgeon's verbal play-by-play is very useful, as he explains what he is doing or discusses his anticipated actions with the resident neurosurgeon. The neurophysiologist can also hear what the surgeon is saying as he calls to the scrub nurse for instruments and describes the patient's state. He hears anesthesia personnel discussing the patient's physiological function. All of this is grist for the neurophysiologists' mill.

Taking the emotional temperature in the operating room means interpreting subtle cues. But this too is crucial to the neurophysiologist's interpretation of the plotted line graphs. Is there a dead silence in the room? Or tension in the surgeon's voice? As one neurophysiologist described it:

> *Neurophysiologist:* What's the feeling in the room? The microphone is very close to the surgeon so I can really get a good feeling for whether he feels like the case is going well or not.
> *Interviewer:* When he is saying something.
> *Neurophysiologist:* Yeah, you can hear it from his voice. You can [also] hear how much activity there is in the room, whether the people are scrambling.

Here the neurophysiologist was listening for the emotional tone of the room as it was evident in people's voices and in the quality of their activity ("whether the people are scrambling"). This information influenced the neurophysiologist's behavior:

> *Neurophysiologist:* Well, if people are agitating, there's a lot going on. I probably would have a much lower threshold for going to the room because I'm alerted then that there's something going on in the room, and that's maybe an opportunity for me to make a significant contribution.

In many cases, the neurophysiologist actually had better access to what was being said when he was monitoring from a remote location than when he was in the operating room. In the operating room, it is sometimes difficult to hear some of what is said because of the noise of equipment and conversations. In contrast, when he is listening to the audio in a remote location, the neurophysiologist gets a clear transmission of what the neurosurgeons and the anesthesia personnel are saying. One neurophysiologist explained:

> *Neurophysiologist:* In fact, the audio is better over the network than it is in the operating room because you can't hear what the surgeons are saying in the operating room. So if you don't know the case, you kind of guess what they're doing. With the audio, you know exactly what they are doing. Because they talk to each other about the steps they are going to take. So you can really anticipate what potentially might happen.

The neurophysiologists also hear other ambient sounds of interest such as the equipment noise, which often provides clues about the progress of the operation or the patient's state. For example, the sounds from the suction device tell everyone in the operating room how much blood is being suctioned. The Doppler ultrasound emits the noise of the heart pumping, the noise of the drill reveals the kind of drill in use, and hammers, saws, and chisels each have their characteristic sounds. Oscilloscopes used by the neurophysiologists and neurotechnicians have audio output that tracks responses produced by stimulation. These responses may go by so quickly that they can't be seen visually on the graphical display, but they can be detected aurally. The audio picks up these sounds.

Taken together, the audio and video in the broadcast facilities provide a much more complete picture of operating room activity than what can be seen in the plotted line graphs alone:

> *Neurophysiologist:* When you look at the computer data by itself [from a remote location], it seems to be one dimensional. When you add the rest of it [audio and video], you get a very rich picture of what's going on [in the operating room].

The use of remote monitoring facilities does not eliminate the need for neurophysiologists to be physically present in the operating room for at least part of the operation. What it does do is allow them to reallocate their time across operating rooms, offices, and conference rooms. The technology gives neurophysiologists more flexibility to move about the hospital on an as-needed basis, rather than being exclusively tied to a small number of operating rooms.

DYSFUNCTION IN THE ECOLOGY

So far so good. But trouble was brewing. During the course of our observations, we encountered a growing concern about *privacy* in relation to the experimental remote monitoring system. We have presented the technology so far strictly from the point of view of the neurophysiologists, for whom it was an unmitigated good.

When we look at the same technology from the point of view of other members of the same information ecology inside the operating room, new facets of the technology come sharply into view. To some people in the operating room, the monitoring system was not simply a wonderful technological support for a new way of working. It was an invasion of privacy. It threatened the sanctity and balance of the social practices that made the difficult work of neurosurgery possible.

After the monitoring technology had been in place for some weeks, we attended a meeting that was convened by hospital staff to discuss rising tensions over the use of the monitoring system. Anesthesiologists, nurses, and neurophysiologists were there. Anesthesiologists and nurses aired grievances about the live broadcast and recording of the audio to remote locations outside the operating room. The concerns expressed were varied. They ranged from worry about malpractice suits to fear of "Big Brother" (a term we heard on several occasions) monitoring job performance.

There was a great deal of confusion over which technologies were actually in use or about to be installed. Some staff felt that the design and installation of the remote audio facilities had taken place in bad faith, without considering the impact on those who did not benefit directly from the technology.

Some of the richness of an information ecology comes from the diversity of its human participants. Ideally, different people in a local ecology work together harmoniously, creating a whole that is more than the sum of its parts. But sometimes, an ecology breaks down because of the very differences in goals and perspectives among its members. Conflict and contradiction emerge. That is what happened in the hospital.

We scheduled a series of interviews with anesthesiologists, nurses, neurotechnicians, neurophysiologists, and neurosurgeons to discuss the privacy issues in detail. The operating room staff had many concerns. The most subtle and worrisome was that the free and unfettered atmosphere in the operating room was being compromised by the remote audio broadcast. The anesthesiologists and nurses argued that by opening up the operating room to people beyond its four walls, the remote broadcast changed the nature of communication *inside* the operating room in significant ways.

They pointed out that both tension and boredom in the operating room are relieved by the relaxed talk and joking that often go on. For example, at a very routine part of an operation Bonnie observed, a resident asked, "Do you remember dead baby jokes?" It was not unusual for people to discuss their weekend plans or poke fun at some of the high-status doctors in the hospital who were not present in the room at the time. Such lightheartedness might seem inappropriate to those outside the immediate situation—such as an attorney or the patient's relatives—but to the staff in the operating room it was a way to cope with an extremely demanding job and a rigid professional hierarchy. The banter and fun in the operating room provided social cohesion in a stressful situation that required meticulous teamwork.

A related concern expressed by the anesthesiologists and nurses was that the new technology might alter the trusting relationship between students and instructors. Students are in the operating room as part of their training, and both they and their instructors might feel inhibited if unseen and possibly unknown listeners had access to their conversation. Students are already nervous enough when they are learning the difficult skills of neurosurgery. A comment such as an instructor remarking "I can't believe you did that" might be open to significant misinterpretation on the part of people who were not present.

The bottom line of these concerns was that the remote monitoring technology would suppress valuable communication in the operating room.

Taking the audio communication out of context was seen as scary:

Anesthesiologist: You can't distinguish between those two [a true problem and something that just sounds bad] on the audio. . . . It can sound terrible and not be, or the opposite. It can sound trivial and be horrible. And you get an incomplete picture without . . . an observer to fill in gaps. . . . What you have is something that could be misconstrued. People are concerned about many things—that real information can be misconstrued, that artifacts and abnormal information can get interpreted as truth, and that truth gets blown out of proportion. So it can be on all those kinds of levels.

There was much discussion of the legal implications of recording video and audio from the operating room. These concerns were valid. In recent court cases, doctors and hospitals had lost because of interpretation (or misinterpretation) of the audio portion of video recordings. The hospital we studied had been recording videotapes of operations for years, at the discretion of the attending neurosurgeon. The tapes went into a library maintained by the neurosurgery department. What drastically changed in the new situation with the remote multimedia system was that anyone who had access to one of the computers on the network could record an operation, so recording was no longer at the discretion of the attending neurosurgeon.

Other concerns seemed to be more an expression of resentment over the way the remote audio and video facilities were installed—without consultation or buy-in from nurses and anesthesiologists. While the neurophysiologists and neurosurgeons were in line to have the technology installed in their offices and conference areas so that they could remotely monitor several operations at once (with neurosurgeons in an advisory role), the nurses and anesthesiologists were not. They felt that the technology did not benefit them, and it was being used at their expense and without their agreement.

The people who installed the system believed that they had informed the nurses and anesthesiologists, but the nurses and anesthesiologists felt they had not. Some of the arguments advanced by the nurses and anesthesiologists against the remote monitoring technology seemed to stem from ill will aroused by their perception that they were being left out of the process. The information ecology had broken down, for social as well as technical reasons. The health of the information ecology was compromised because the varied interests of the different players in the ecology were not taken into account as the monitoring system evolved.

The nurses did not stand to benefit from the remote monitoring facilities in any way because of the nature of their jobs. They were at the lower end of the status hierarchy in the operating room, and they were the ones who worried about the system being used in a Big Brother capacity to monitor their job performance.

This fear was not unfounded. The neurosurgeons were considering the use of a wide angle or "environmental" camera that would show the operating room itself and the staff as they worked. This bit of knowledge came out as gossip rather than in a formal statement to the staff. At the meeting on privacy issues it was stated that "no one assumed an environmental camera would be a problem." The rationale for a wide-angle camera was given vaguely as "understanding the gestalt of the case."

But an environmental camera is easily seen by almost any worker as a potential threat. The idea of installing a camera with the ability to watch and record people as they work is clearly one that needs to be openly discussed. The offhand rationale invoking the "gestalt of the case" failed to explain the potential value of a wide-angle camera. It did nothing to quell or even acknowledge understandable fears.

Another argument that was raised against the use of the remote broadcast was that it would invade the patient's privacy. However, patients essentially sign away all of their rights to privacy in this setting, so this concern seemed somewhat manufactured. Individual doctors may go to great lengths to assure patient privacy, but legally, patients sign consent forms that remove their rights. The video recording that had been going on for years also reduced patient privacy, yet patient privacy did not become an issue until the remote monitoring facilities were installed.

The anesthesiologists had an interesting position in the privacy controversy. They did not stand to gain directly from the current system, yet they could see its usefulness. They too wanted to have remote facilities for monitoring physiological responses, but they were not as far along in a separate development effort to create such a system for themselves. Although they complained about legal dangers and other issues, they also offered ideas for privacy safeguards for future systems.

Many staff were genuinely puzzled about why the remote audio broadcast was useful. There was confusion about when the remote facilities were active, who might be listening at a remote computer, when a recording would be made, whether both audio and video were recorded, when new facilities would be installed, and what they would be. Some of the nurses were not aware that their voices had been recorded on videotape for years, using the earlier video capture system in the operating room. The installation of the new remote monitoring system significantly raised people's awareness and concern about capabilities that had been in the hospital for some time, as well as concerns about the new system.

With the introduction of the new monitoring system, the information ecology experienced a severe and wrenching realignment. Information was taken out of its original context and presented in a new context, without the buy-in of the people who generated the information.

Information changes shape and function dramatically when its broadcast boundaries are altered. The context of information presentation and dissemination in the operating room was suddenly, and for some, painfully, transformed. A remote broadcast inevitably reduces individual control over the kinds of information that people can consider private or personal. In the operating room, what was once an ephemeral event in a small, well-defined, visible space, with known participants, became a situation in which speech and action could be permanently recorded and broadcast live to remote, unseen, and possibly unknown viewers and listeners. The tensions over the system created a disturbing dislocation in the information ecology. Many participants in the ecology had no opportunity to offer their opinions on the technology nor their suggestions on how to evaluate and improve the design of the new system.

It is important to remember that an information ecology, unlike a biological ecology, is *designed*. A new technology such as the remote monitoring system is not a blind fact of natural selection but a product of the human head, hand, and heart. As such, it is the responsibility and privilege of people in the local ecology to shape new technologies and practices. In the hospital, the dysfunction in the information ecology came about for two reasons, which could have been anticipated and dealt with earlier.

First, the system did not contain simple privacy safeguards. For example, there could have been an indicator showing when the video and audio were being recorded or broadcast. Second, and more important in our view, there was no systematic and thorough effort to inform the staff of the benefits and features of the technology and to involve them in decisions about its use.

ASKING STRATEGIC QUESTIONS

We have suggested that people routinely ask strategic questions when they are planning to change their information technologies. In the hospital, it would have been good to ask some of the most basic strategic questions, to share perspectives on why the system would be useful, discuss the key characteristics of the technology, and understand people's feelings about what makes teamwork in the operating room successful and enjoyable.

Strategic questions about consequences would have been especially helpful. Many concerns would have arisen early in a discussion about possible consequences, including the change in who controls video recordings, the impact on student-teacher relationships, and changing patterns of informal communication. The privacy issues could have been discussed in an atmosphere of cooperation rather than conflict.

A fruitful dialogue began informally and spontaneously at the privacy meeting. The anesthesiologists offered several suggestions for privacy safeguards, such as showing one kind of "on-air" light when an operation was being recorded and another kind of light when the remote audio was being broadcast. It was suggested that the lights be placed not only in the operating room itself, but also in the two hallways that have entrances into the operating room, so people could mentally prepare themselves before entering the operating room.

Suggestions of this kind can make a big difference to the healthy functioning of an information ecology. They are the hallmarks of creative coevolution of technology and work practice. Such suggestions can only come from participants in the ecology themselves, because they know the details of a particular situation. In the hospital, for example, it was important to think about details like the layout of the operating room and the need to be prepared before entering such a demanding work environment.

Through the process of questioning, commenting on, and contributing to the design of a new system, people come to understand it more fully and feel less threatened by it. A more inclusive process allows questions and ideas to be taken into account before they lead to the kind of tense situation we encountered in the hospital.

As we have been arguing throughout the book, it is up to all of us to participate in our own information ecologies. We need to carefully consider new technologies to make sure they will fit well with our work practices and values. People in the hospital did end up contributing their ideas about the remote monitoring system, but they did it in a reactive way. Asking the strategic questions throughout the process would have prevented some of the tension and dislocation in the hospital information ecology. Broader participation can lead to better system designs, while keeping the information ecology healthy and functioning throughout times of major change.

12

Diversity on the Internet

In the earlier empirical chapters, we discussed examples of information ecologies we have encountered in our research studies. In this chapter, instead of an in-depth look at one particular corner of the world, we offer a reflection on the Internet as a riveting global phenomenon with important implications for local information ecologies.

There is no single Internet information ecology. Information ecologies are local habitations with recognizable participants and practices. Nothing as huge as the Internet can be seen in that light. We view the Internet as a set of environmental conditions that provide a substrate for the growth of ecologies that span traditional geographic or social boundaries. The Internet can serve as connective tissue between and within local information ecologies.

The Internet, in the broadest sense, is a network and a set of communication protocols. It allows computers in different locations to communicate and exchange information when they are connected over telephone wires or wireless modems. The most common way for home computer owners to connect to the Internet is using an ordinary phone line and a modem in their computers.

Typically, home computer owners sign up with an Internet service provider, which charges a monthly or hourly fee for the service. The service provider makes available to its subscribers certain telephone numbers for their computers to dial, and once the subscribers are connected to the service provider, they can reach the Internet. There are hundreds of such service providers in some areas and only a few in other areas. Students can usually access the Internet through their colleges and universities, and some employees have access through their companies.

Computer users do many things with this communication capability. They send electronic mail messages to one another. They join group discussions on "listservs" or other kinds of mailing lists. When people are on a mailing list, they can send an email message to everyone else who has signed up for the list. People monitor "newsgroups," or bulletin boards, on topics of interest to them, posting their own messages when they have something to add. They participate in live "chats," which are like multiparty phone conversations that use typed-in text instead of voice communication. Internet users play in collaborative virtual worlds like Pueblo. They find information and use services on the World Wide Web, a vast, searchable collection of documents called "Web pages." People also publish their own information and services on the Web.

What deeply interests us about the Internet is its potential to foster diversity. The Internet is global in reach but decentralized in structure. It is accessible to individuals and small local groups as well as large organizations. Communication and information sharing can take place between one person and another, among limited groups of people, or throughout the whole world of Internet users. On the Web, you can produce and distribute information and interactive applications to a broad audience. You can also find and use what others have created. The Web is not a one-way channel for carefully controlled content, as with mass media or government publications, but an enormously varied tapestry created by many thousands of people.

We will describe some of the different kinds of encounters people can have on the Internet today, which are notable for their variability and richness. Some of our examples focus on information ecologies in which the Internet plays an important role, and others simply show off some of the information and services available on the Internet. Through these examples, we want to highlight the diversity inherent in today's Internet. After giving a flavor of the immense variety of the Internet's information and services, we discuss threats to diversity on the Internet. The Internet hasn't settled down yet, and its future evolution is uncertain. We hope the Internet will not go the way of a bland interstate but will retain its current Byzantine charm of virtual crooked streets and alleys, cyberplazas, and electronic corners and crannies.

WHAT'S OUT THERE ON THE INTERNET?

Several years ago, Vicki gave an introductory workshop on the World Wide Web to a group of high school teachers. She had the impossible task of explaining the then-new phenomenon of the World Wide Web in a room with no live Internet connection. She decided to prepare ahead of time by taking several successive screen snapshots of a Web session, to show the teachers what it was like to search for something on the Web and browse the results.

Picking two words out of the air, she used a popular search engine to search first on "Mexico" and then on "fish." The first item in the list of search results for Mexico was the entire text of the then-unratified North American Free Trade Agreement (NAFTA). The first item in the fish list was a page about reef tank aquariums, assembled by a passionate home expert, containing pictures and videos of reef tanks, sources for mail-order supplies, detailed directions on how to put together a reef tank, and lists of questions frequently asked by fish lovers.

These two examples nicely illustrated the breadth of information on the Web, ranging from official to unofficial, authoritative to amateur, textual to multimedia, fixed to changing, single document to a compendium of material from different sources. The teachers began to fathom this new technology, with some bafflement because of its newness, but also with strong glimmers of interest.

Everyone has their own favorite information examples on the Web. A simple but useful store of information is zip codes—you can find a zip code anywhere in the United States just by typing in a street address on the right Web page. This is perfectly straightforward information, with nothing very exciting about it, but in paper form it is scattered in post office publications and telephone books that are organized by (and in the case of phone books, only readily accessible in) different local regions. Having a single point of access to scattered local resources makes a big difference.

Another example is street maps. By connecting to certain Web sites, you can generate a street map around a particular address of interest to you. This is more customized than the zip code information; the map may actually be different for two nearby addresses. But it is useful in the same way, allowing people to find local information from inside *or* outside that local region.

Information is what we typically think of when we think of the Internet, but there are also other crucial connections enabled by the Internet. We see four basic applications of Internet communication: connecting people with *information, services, goods,* and *other people.*

Connecting People with Information

The Web has something in common with the world of eighteenth-century print shops, where local printers produced broadsides and pamphlets in small runs, to be distributed today and replaced by something new next week. Today, most print publishing is focused on large-scale production, and the means of production are available only to a relatively few writers whose works are expected to justify the expense of publication and distribution. Though the Internet reaches millions, it has the lively and haphazard feel of an earlier information era. Print publications such as *USA Today, Time,* and now even the venerable *New York Times* have converged stylistically and in other ways, but there is an amazing variability in the "personalities" of Web sites. Some are slick and glitzy. Others are spartan. Some have drop-dead gorgeous graphics, and others offer just-the-plain-text facts. You never know what you'll find on the Internet.

The up-to-the-minute feel of the information on the Internet is exemplified in its sports coverage. Sports reports can be updated on television or radio too, but this is transitory information—you have to be tuned in at the moment the information is broadcast to hear it. On the Web, you have consistent easy access to the very latest game schedules, background information on teams and players, and up-to-date scores. You can gaze at them as long as you wish, save them on your computer, and print them out.

Air fares, like sporting event scores, change often. Unlike sporting events, air fares are a highly decentralized form of information, scattered among different airlines. Internet sites have sprung up to provide sophisticated trip-planning capabilities that use the dispersed fare information. If you are planning a trip to Hawaii, you can specify where you're leaving from, when you'd like to go, what times of day work for you, which airlines you prefer because you carry frequent flyer miles, and so on, and then you can see a list of flights with available seats in your price range. If the fares are too high, you can arrange for email to be sent to you whenever prices drop.

The Internet was perfect for exchanging information about the discovery of the Shoemaker-Levy comet that plunged into Jupiter in July 1994. This information was new (certainly not yet in books—discoveries like this only make it into books long afterward), it was changing rapidly, it was excitingly visual, and it was scattered all over the world, as images and other data were captured from different telescopes and observation stations. Very quickly, Shoemaker-Levy home pages were put up on the World Wide Web to bring together information provided by different teams of astronomers.

It was an extraordinary experience for the general public to be able to look into a new discovery as it was unfolding, as it was being discussed and probed by scientists. The thousands of people who visited these Web sites every day certainly could not have squeezed into a corner of the Jet Propulsion Lab (one of the sponsors of a Web site for the comet). But these people's presence as observers and even participants online was exciting and valuable. It is good for the practice of science to have more people aware of its hard work and new ideas.

FAQs are an enduringly popular form of information on the Internet. FAQs are lists of "Frequently Asked Questions" (along with their answers!). In the early days of the Internet, FAQs abounded in technical areas such as programming and networking. But the idea is simple and elegant, and it transfers easily to any area of knowledge. A FAQ is the solution to a common problem. On a mailing list or newsgroup, certain questions are raised over and over again, as new people "tune in" for the first time. Instead of snarling at the new people for asking one more time a basic question that the old-timers are heartily tired of hearing, newcomers can be pointed politely to a FAQ: please read the FAQ first, and if you have further questions, feel free to ask the whole group.

The contents of FAQs are unique to the topic. In a parenting newsgroup, the common questions are about getting your child to sleep through the night (when? how?), breastfeeding, circumcision, and good books and toys. In a gardening newsgroup, the questions are about pest control and favorite hardy fruit varietals. We think FAQs are interesting because they show the value that personally crafted, unofficial information can have. FAQs are a volunteer labor, and they do not come from any certified source. They are assembled from the collective wisdom of a large and possibly anonymous group of people, yet they are usually both succinct and extremely useful.

It is this kind of individually crafted, personal publication that makes the Internet unique. Everyone is passionate and expert about something (just like that reef tank aquarium enthusiast). It is a good thing for people to be able to share their expertise.

Connecting People with Services

The distinction between information and services is sometimes blurred. A FAQ, for example, is both. It may look more like information, because it is in the form of a document in question-and-answer format, which is easy to look at on a screen or printed page. But it is also a service, provided by the FAQ keeper of the moment in each domain of knowledge for which FAQs are kept. The Internet metaphors initially centered around providing access to information as a fairly inert substance, but increasingly, we are seeing that information is much more useful on the Internet when it is provided as part of a service.

In the airline example above, we classified the ticket data as information because in the air travel Web sites we've used, we have had to jot down our choices and then get on the telephone and call the airline. When you take the extra step of reserving your tickets, as you can on some sites, then we would put this example into the service category.

Some services are official, sponsored by recognized institutions such as the U.S. Government. It is now possible to order a passport online through a government passport service. You can find tax forms online at tax time. Online college applications are available. Just as with some of the information examples above, the "sweet spot" of the Internet that we notice here is a major shift in the visibility of information. Sometimes scattered information (such as college applications) is pulled together into one well-known location. Sometimes information that is usually in only a *few* fixed physical places (such as tax forms or passport applications) is presented in more broadly accessible forums.

Equally useful and often more interesting and creative services are those that individuals have developed on their own. For example, there are fix-it sites on the Web, where expert plumbers and cable television installers provide step-by-step instructions (with pictures!) for common repair or installation tasks. At some sites, the experts are even available online, for free, to answer questions on home repair from befuddled amateurs.

We do not think that people are freely providing these services only because electronic currency exchange is not yet mature. We think that many people *want* to share what they can do and what they know best, because they like doing it and feel rewarded by it. The Internet provides a huge sort of piazza for all sorts of exchanges and offerings, including many that don't involve money.

Some services do cost money, of course. Traditional service industries such as investment management have successfully expanded to the Web. You can investigate and trade in your mutual funds online. You can take courses online, using a combination of traditional and untraditional materials and discussion formats, including books, CD-ROMs with lectures on video, and online chat rooms to talk with other students, who may be anywhere in the world.

Services on the Internet are still fledgling. Certainly the availability of richer media for communication and sharing information will enlarge the possibilities. With text as the most convenient medium for information in a global network, translation services seem an obvious avenue to pursue—and indeed, they are already available. But as network speeds to the home improve, and video, audio, and animated graphical images become more practical, a different generation of services will arise.

Connecting People with Goods

Let's look at some interesting examples of Internet commerce that incorporate the diversity and heart we have championed for information ecologies. One example of a consumer-oriented business on the Internet is a book purchasing site that is widely known by its Web address, Amazon.com.

We do not know whether Amazon.com's claim to be "Earth's Biggest Bookstore" is true, but with two and a half million books available at the time we write this, we would not be surprised. Still, it is not just access to millions of books that recommends Amazon.com to us (though we rarely visit the site without finding something we'd like to buy).

What really makes Amazon.com interesting is the number of ways it provides for book buyers to find and evaluate books. Visitors to the site can search for a particular book they're looking for, or browse the virtual aisles of children's books, art books, science fiction, poetry, nature, cookbooks—whatever you might expect to find in a physical bookstore. You can look for books that have won awards, books on the *New York Times* best-seller list, books that are the current "editors' choices," or books that are currently the most popular among Amazon.com shoppers. Visitors can sign up to receive email notices about new books on particular topics. If you want an out-of-print book, Amazon.com will try to find a used copy for you.

As you browse through descriptions of books (complete with pictures of the dust jackets), you can read the comments other readers or the books' authors have submitted—a feature that takes advantage of the fact that the Web is a two-way information exchange, not just a broadcast medium. The site's diversity and the friendly attitude it projects give a sense of having a large, yet personal, bookstore at your fingertips. It is a far cry from interacting with untrained clerks in a sterile shopping mall atmosphere.

Amazon.com is a real alternative to the experience of shopping in a large chain bookstore. How many chains will try to find a used copy of an out-of-print book for you? None provide the kind of evaluative material available at Amazon.com. Most important, the chains stock only the best-selling books, so shopping there is like eating at McDonald's. You know what exactly what you are going to get. Amazon.com, by contrast, has practically everything. It's a giant bazaar—diverse, full of serendipity, with endless alleys and mazes to explore.

Another example of Internet commerce that offers consumers choice, control, and variety is the Virtual Vineyards wine shop on the Web. Like

Amazon.com, Virtual Vineyards has a friendly, outgoing image. It offers wines selected by the "proprietors" and accompanied by descriptions that go beyond the numerical ratings on grocery store shelves. The site also includes wine sampler packages, information and selections from small artisan vineyards, advice (and menus) on matching food with wine, a monthly wine program, and a question and answer column on wine.

Both Virtual Vineyards and Amazon.com have found ways to take advantage of some of the special attributes of the Internet: the ability to communicate *in both directions* between buyers and sellers, update the inventory or presentation of goods frequently, provide specialized services in a cost-effective way, and respond to changing conditions and shopper interests. The designs of these two businesses emphasize the natural strengths and affordances of the technological medium. They still borrow from the metaphor of physical shopping where possible (such as a button on the screen to add an item to your "shopping basket") to make the experience easier for new users.

Both of these commercial sites also show the value of the mediation provided by the middleman. Shoppers here are *not* going directly to publishers or vineyards. Instead, they are using the selection, organization, information, and extra services provided by the people who manage these sites. There is a paradox here: though Internet businesses may be able to carry more goods and handle many more customers than traditional storefront businesses, the best sites feel more like neighborhood stores than giant warehouses. They are idiosyncratic and interesting, not bland and homogenized. A windowless Wal-mart, "real" though it is, is a far less engaging place to be than Amazon.com or Virtual Vineyards or any of thousands of other well-designed and well-maintained Web sites.

Connecting People with People

The Internet hosts pure people-to-people communication in the form of email, chats, newsgroups, and other virtual places where people do everything from fall in love to debate political issues to emote about events such as Diana Spencer's death. We don't have any idea of how to quantify the immensity of the communications that take place on the Internet, but even the most casual stroll through cyberspace will give a flavor of the diversity and breadth of the exchanges.

We find this diversity particularly important in an ever-more packaged world where Ellulian standardization constantly hovers. While much of the communication on the Internet is humdrum and unremarkable, the fact that it is just people communicating their own thoughts to other people is heartening. Not every human interaction has to meet a high intellectual standard. Nor should it be packaged through the filter of focus groups, survey research, and other marketing tools that transform varied responses into flat, codifiable indices. On the Internet, real semiotic activity, from the ridiculous to the sublime, flourishes as an everyday occurrence.

A fascinating experiment in using the Internet for people-to-people connection was conducted by anthropologist Melinda Bier and her colleagues in Florida. They outfitted six low-income families with everything needed to use the Internet, including training and technical support.[1] The families had Internet access for over a year. During that time, every family made heavy use of the Internet and the researchers found that Internet usage "enabled powerful emotional and psychological transformations on the part of the research participants." Study participants used the Internet to access medical information, to join support groups, investigate scholarships, make local transportation arrangements, exchange recipes, and help their children with their homework. (They also attended the O.J. trial and engaged in online pie fights.) They learned to type. They began to use the computer for other functions such as budget management and personal correspondence.

One participant reported a use of the Internet that had been particularly important to her:

> I was able to get information about my niece's disease. I wasn't even able to pronounce the word, but I was able to email the librarian at the University of Iowa, the radiology one, straight to her and then she directed me how to get more information. You go into the virtual hospital. You know if you had to do it by yourself, by phone, forget it, you'd give up—ha—like a slow boat to China. Now you say, "OK, I can pull it up on the Internet and we'll find out right quick."

Another participant joined a support group for diabetics and learned more about the importance of diet and exercise than she had from her doctor. In addition, she received emotional support from her group.

Others helped their children with their homework and became more involved in the schools. One participant found information about federal grant monies for schools and sent it through email to the vice principal of the school. Some of the parents were eager to share their new skills and volunteered to serve as technology trainers for their children's teachers.

Many experienced a new sense of themselves:

> It doesn't matter that I'm on SSI [Supplemental Security Income] and living in HUD housing. On the Internet I can sit down with people making $150,000 a year or more and I can keep up with their conversation. There's no way I thought this would be possible.

An interesting aspect of this study was that the study participants did not fit the Internet stereotype of the affluent white male user. Some say that the Internet was designed by and for young, middle-class white males. The implication is that the Internet does not meet the needs of women, people of color, senior citizens, or the economically disadvantaged. But the Florida residents embraced the Internet, using it in myriad and creative ways.

The results of the Florida study show that the Internet can serve many different kinds of people well. It is significant that Bier and her colleagues provided equipment and technical support to the study participants. It may be not so much that white males have a technology designed to fit their cultural profile, but that they have the money and technical expertise to adopt the technology and use it fruitfully. When economic and educational issues are dealt with, the Internet is no more culturally biased than the telephone.

At the end of the thirteen months during which the study participants had Internet access, a sad and infuriating reality awaited them. The computer equipment, which was all on loan, had to go back to the researchers' institution. Because Internet use had become an integral part of their lives, study participants were understandably angry. One said, "Oh yeah, you university people are worse than the dope peddlers; now that we're hooked, you're going to take it away." The researchers summarized, "Participants fear losing their newly acquired sense of identity, education, and community." The researchers extended the loan period of the computers as they searched for alternatives to removing the machines altogether.

In Oakland, California, a large city with many low-income residents, a program is underway to teach job skills to welfare recipients in their homes, through Internet connections.[2] The multimillion-dollar program is funded by the city of Oakland and the California Wellness Foundation. IBM will provide computers at a discount. Training will take place at a center within the housing project, and then residents will have their own terminals in their homes, which they can use once they have mastered basic skills. Training will include basic literacy, office skills, word processing, spreadsheets, and database management—skills needed in the marketplace—noted the program's sponsors. Children will be able to use the Internet for their homework.

In our discussion of digital photography in school, we noted that access equity is a value that can be promoted in school. Some commentators believe that computers are not essential in the classroom—that "service jobs" do not require them. The Oakland project shows the kind of catchup we have to play when students do not get basic computer experience in school. A nonprofit group, Computers in Our Future, reports that more than half of new jobs require some sort of computer skills. They note that only 14 percent of low-income high school students have access to computers, while 82 percent of students in more affluent homes have access.[3]

A challenge for our society is to provide universal access to the Internet for those who cannot afford it, just as we do with telephone service. (Check your bill; you pay a few cents a month for someone else's phone.) A U.S. Department of Commerce report declared that "many of those most disadvantaged in terms of absolute computer and modem penetration are among the most enthusiastic users of online services that facilitate economic uplift and empowerment."[4] Bier's research demonstrates that these words are not mere rhetoric; there is real potential for the Internet to serve all segments of the population. The government should continue to play an active role in furthering Internet access, as it is trying to do. As citizens, we should make sure that our government does not back off and let market forces alone design the global network that the Internet will become as it matures. We successfully promoted people-to-people communication access with the telephone, and we can do it with the Internet.

A PERSONAL VIGNETTE

We have given many examples of how the Internet can be used by different people, but let's turn now to a different kind of example. To see what the Internet looks like from just one person's point of view, we asked one of our children to describe what he does with it. We find it especially interesting to see what the Internet looks like from the perspective of a young person who is growing up with it.

Anthony is a sixteen-year-old high school student. He and a group of his fellow students are using the Web extensively to research a group report on the literature of the Beat poets for their English class. They have found many useful sites, such as the Beat Generation Archives and Kerouac Quarterly, which are monitored and updated by scholars, historians, and fans of the beats. The students have also visited the San Francisco Public Library and City Lights Bookstore to gather information. They use the Internet as one resource among many, not as a replacement for other resources.

However, there are some features of the Internet that make it a unique resource for Anthony and his group. They especially like the fact that the sites they use are constantly updated. They also value the two-way communication. The editor of *Kerouac Quarterly,* an online journal about the Beats, started a chat room called Kerochat. The students have used the chat room to talk the editor himself and some of the people who actually knew the Beats.

Anthony also participates with a set of school friends in a "wwwboard," a mechanism similar to newsgroups but somewhat easier to set up. Their board constitutes a thriving information ecology, with frequent and varied postings. Any member on the board can post anything, and, as Anthony noted, "We learn all sorts of stuff about each other from philosophy and views on ethical and moral issues to the type of humor that they like." Of course, giving teenagers yet another communication channel might not seem to be forward progress, but each medium has its own shaping affordances. Anthony pointed out that "some things are better said in writing and this is an easy way to put written stuff where everyone will read it."

Some recent "threads" or lines of discussion in the wwwboard include the posting of "The Love Song of J. Alfred Prufrock," for the group to read and respond to, "Funny video editing," an email humor post from another board, and "Male vs. female perspective," advertised as "a philosophical debate."

Anthony's use of the Internet raises some questions. How will he know whether the information from "people who knew the Beats" is any good? How will potentially divisive topics such as "Male vs. female perspective" be handled by young people who may not have much experience in working through their misunderstandings in a public forum?[5] Is hanging out in Kerochat a good use of his time?

We believe that part of the answer lies in coaching students on how to participate more effectively in information ecologies. As we were working on this book, we got an email message from Frances Jacobson, a reference librarian who teaches a high school class on how to use the Internet. Her message shows what it will take to provide this coaching:

> I'm up to my ears in the day to day stuff of kids. We're really struggling with the new configuration of our introductory computer science course. Last week's lesson was newsgroups. I posted 6 hypothetical scenarios of netiquette quandaries for the students to respond to. Between helping them negotiate net news software (ugh), reading their responses, and fighting with the technology, we're going a little crazy. I'm also grading guides they wrote on using various encyclopedia titles in different formats. But when I'm not feeling overwhelmed, I'm enjoying it. Really.
>
> Cheers, Frances

Jacobson is helping her students forge new relationships with technology. She is giving students technical skills, reading what they write, and evaluating their use of information resources. As a reference librarian, she can provide tips and techniques on how to conduct more effective searches and help students learn to sniff out good information from bad. Moreover, Jacobson is socializing the students into being courteous users of the Internet by teaching them "netiquette." We believe that this kind of explicit training on how to use the Internet productively and courteously will make it an immeasurably more valuable resource in the future.

DIVERSITY AND THE INTERNET

Jacques Ellul predicted that we were heading for an increasing sameness in our institutions, practices, and ideas. He felt that the dynamics of *technique,* with its ruthless obsession with efficiency, would drive out aesthetics, ethics, fantasy—sources of diversity that work against the mechanical efficiency of pure technique. We see Internet activity as a counterweight against the uniformity and standardization that Ellul foresaw. Access to media for expression are vitally important in the age of information, and the Internet makes new means of expression broadly available.

At the time of this writing, more than 170 countries have Internet access. As one powerful example of how the Internet supports diversity, we can look at the way it is being used to help protect and promote ethnic minorities in countries all over the world.[6] Cultures under threat— Mayan Indians in Guatemala, the Kuna of Panama, the Aymara in Bolivia and Peru, Burmese in exile, the Sami of Finland, and many Native American nations in the United States and Canada, to name a few—use the Web, email, newsgroups, and discussion lists to further their interests.[7] These media allow for exchanges of information that support cultural identity, help with practical affairs such as politics and health care, and enable the preservation of ethnic languages.

The example of ethnic minorities using the Internet to promote their interests is particularly compelling to us because they are using technology in the context of their own local values. Many of the examples we discuss in the following paragraphs involve distinct information ecologies. Others involve more diffuse documentation and broadcast of information about the activities of large ethnic groups and their relations to wider cultures, such as the Kuna's participation in international discussions of environmental issues.

In Guatemala, the Maya are "working to retrieve all the information that pertains to their culture starting with programs of linguistic restoration, as well as documents that may shed light on the legitimacy of their ancient territorial claims," note anthropologists Guillermo Delgado and Marc Becker.[8] Delgado and Becker describe how the Kuna of Panama use the Internet to promote environmental issues of importance to them. Some Kuna have become leading figures in debates over biodiversity, indigenous property rights, and DNA collection. Delgado and Becker observe that the Kuna's control of electronic media has been an important aspect of their visibility in international discussions.

A number of different Native American groups live in the Northwest Territories of Canada. They maintain many Web sites. Surfing these sites gives an instant sense of the character of life in the far north. One of the many interesting pages is sponsored by a group called Pauktuutit, which represents Canadian Inuit women. Pauktuutit is involved in a number of social issues, including midwifery projects, alleviating problems of substance abuse, and promoting housing and economic development. Among other things, Pauktuutit has coordinated an antismoking campaign, held an Inuit fashion show, and published a paper called "Inuit Women—The Housing Crisis and Violence" for the Canada Mortgage and Housing Corporation. Other Web pages in the Northwest Territories discuss educational opportunities, employment, native culture, land rights, government, tourism, and many other topics.

Barry Zellen, executive director of the Native Communications Society of the western Northwest Territories, observes that most Native American northerners do not yet have access to the Internet, because most do not have computers and modems.[9] But efforts are under way to wire the north and to provide access to as many as possible. For example, telecenters are being planned in regions where most homes lack computers. Margaret Gorman, executive director of the Denendeh Development Corporation, says that issues of concern are "ensuring that the system is used by the people of the communities; training local people to maintain the system; reducing the threats of negative influences like cyberporn and online gambling; developing northern aboriginal content on the Internet; and identifying ways to use the Internet to serve the communities' needs such as marketing traditional arts and crafts." Gorman's articulation of native concerns is an excellent example of local values being applied to the thoughtful planning of technology use.

In Burma (Myanmar), we find a very different situation than in the about-to-be-wired North. For all practical purposes, the Internet is banned in Burma. Every fax machine and modem must be registered with the government—a repressive military junta that controls all media. Anyone found in possession of an unregistered fax machine or modem can be imprisoned for up to fifteen years.[10] A very few people, such as members of NGOs (nongovernmental organizations), international businesses, and some government officials, can receive and send email once a day. Otherwise, there is no Internet in Burma.

But as anthropologist Christina Fink observes, Burmese exiles outside the country "have been quick to take advantage of email and the Internet, both to distribute information in a timely fashion and to organize resistance activities." Burmese in Japan, Thailand, and Western countries use the Internet to stay in touch with each other and to educate people about conditions in Burma. Some exiles have home pages with photographs and information about the junta's use of forced labor and relocation campaigns. Resistance groups such as the Karen National Union regularly post news on discussion lists, including reports of human rights abuses within Burma. Information about the mass exterminations of Shan villagers that took place in June and July 1997 was broadcast on the Internet—the only way for this information to find its way outside local areas of Burma. Fink notes, "This news is essential because it is virtually impossible for journalists to travel to sensitive areas in Burma, particularly the ethnic minority homelands in the border regions."

Many of these reports are translated into English and emailed around the world. And some of this news carried on the Internet is then broadcast back into Burma by radio stations based outside the country, such as the British Broadcasting Company, Radio Free Asia, the Democratic Voice of Burma, and Voice of America. Fink provides many other examples of the uses of the Internet by Burmese exiles, and remarks, "Because Internet use is comparatively cheap and fast, it has provided the ideal space for mobilizing people who are geographically separated by thousands of miles."[11]

The Burmese case is an interesting counterpoint to the pessimism of some technology commentators. Jerry Mander, an incisive critic who developed important arguments against television, wrote about Burma in 1991 in his book *In the Absence of the Sacred.*[12] One of his overall arguments was that computers mostly help large corporations and governments—but do little for everyone else.[13] In 1991, Mander noted that American newspapers reported almost nothing about Burma and that the political situation seemed to be at a stalemate, with the military firmly entrenched. His short, terse paragraph on Burma left the reader with the sense that nothing could be done.

Much has changed since 1991 with respect to the Internet. Yet Mander's later writings, such as *The Case Against the Global Economy,* suggest that he has not changed his mind about computer-based technologies.[14] For example, he wrote, "The new telecommunications technologies assist the corporate, centralized . . . enterprise . . . far more efficiently than the decentralized, local, community-based interests, which suffer a net loss."[15] But we think the activities of the Burmese activists in exile tell a much more nuanced and optimistic story. Not only is material being documented and broadcast that cannot be touched by American (and other) journalists, for the reasons Fink explained, but there is also a communication channel of great practical utility being applied to the active resistance of oppression.

Mander's high-level argument does not take into account some of the on-the-ground activity that gives cause for optimism. A lot is missed when we do not look empirically at specific uses of specific technologies. We have seen this blind spot with other technology critics such as Larry Cuban, who dismisses technology in schools, and Clifford Stoll, who believes the Internet is snake oil. Looking at real human activity at local levels often tells a different—and more empowering—story than looking at social forces in the large. We find inspiration in the activities of Burmese exiles fighting the junta, Kuna promoting environmentalism, Maya defending their land and culture, and Inuit women taking on issues of health and housing—all aided by the Internet.

THREATS TO THE INTERNET

Right now the Internet is immensely diverse and open. But it may not stay this way. The current openness is threatened by the possibility that powerful economic interests will commercialize the Internet to the point where it will completely change in character. There are strong economic incentives to mold the Internet into a particular kind of place that emphasizes consumption and commerce, rather than sustaining it as the many diverse places it is now. In the future, it may be too expensive to post our own Web pages, we may find only commercial sites, we may be bombarded with ads in every Internet interaction, and human mediation may be replaced by automation.

We are not against the addition of commercial interests to the Internet. There are many interesting and valuable Web sites run by companies to increase their profits. Such sites add to the diversity and utility of the Web. Even when sites charge for access to information or services (as the *New York Times* is considering for its news archives), we still feel that we are better off than when these resources were available only to a much smaller number of high-paying corporate customers. What concerns us is the prospect that commercial ventures might overwhelm noncommercial and individual uses of the Internet.

Some people envision the Internet as a place that is *primarily* for commerce. Bill Gates, the CEO of Microsoft Corporation and one of the most influential figures in the development of computer-based technology, has written a book called *The Road Ahead* in which he cogently sketches this vision.[16] Gates's powerful position makes it necessary to pay serious attention to his projections about future uses of the Internet.

Very early in his book, Gates calls the Internet the "ultimate market." There are many metaphors used for the Internet, most involving information, communication, and connection, but Gates hopes to make the Internet a place devoted to commerce. The user of this Internet is cast as the ultimate consumer. This theme is explored is his chapter titled "Race for the Gold," which is meant to invoke the fever of the California gold rush.

The Road Ahead is packed with vignettes about consumption in the "global information market." A central feature is that there are no pesky, expensive middlemen standing between consumers who purchase goods and producers who make them. Stores, salesclerks, and agents such as insurance salespeople and stockbrokers will be rare and quaint.

Gates calls the notion of direct purchasing from producers *friction-free capitalism*. Of course, someone will have to collect the fee for the Internet transaction that brings the consumer and producer together. This harmless little service, not really a middleman at all, will naturally go to some software provider—whoever wins the race for the gold.

Gates's vision of information access on the Internet has a strong commercial emphasis. He declares, "As the Internet is changing the way we pay for communication, it may also change how we pay for information. . . . I believe the most attractive information, whether Hollywood movies or encyclopedic databases, will continue to be produced with profit in mind."[17]

In Gates's view, television—with its expensively produced information structured by commercial interruption and broadcast to huge audiences—is a model of how the Internet should evolve. Gates expects that schools will soon turn to "entertainment-quality interactive learning materials." The Lightspan Partnership, he notes, "is using Hollywood talent to create live action and animated programs. . . . These programs will be available on televisions in homes and community centers as well as in classrooms."[18] Educational content will be delivered in a television format over the Internet. Instead of children using the Internet to create electronic journals, visualize molecules, or run their own show in a simulated stock market, to name a few possibilities, they will be watching television at school.

In Gates's vision, entertainment, education, and consumption blend into one big lucrative soup. On the Internet, "interactivity will marry convenience with entertainment."[19] For example, film will become a form of shopping, as brand-name product appearances turn into interactive shopping experiences. Gates observes, "If you are watching the movie *Top Gun* and think Tom Cruise's aviator sunglasses look really cool, you'll be able to pause the movie and learn about the glasses or even buy them on the spot—if the film has been tagged with commercial information."[20] We cannot even speculate on what Fritz Lang, who cared very deeply about the *ideas* in his films, would make of watching a movie to buy sunglasses. Think of your favorite films. Would watching them for shopping opportunities have enhanced their meaning or entertainment value?

For Gates, consuming on the Internet will take many forms. We were particularly chilled by Gates's casual mention of gambling as an appropriate use of the Internet. He points out that "the profits garnered by casinos are incredible."[21] While this book is not the place to enumerate the social dislocations caused by gambling, it is well known that gambling causes hardship for addicts and their families. "The highway will make gambling far more difficult to control than it is today," Gates observes without alarm.[22] Do we really want to use the riches of the Internet to make it easier to bring online casinos to millions of households?

In *The Road Ahead,* Gates acknowledges potential problems, but he always has the same answer: the benefits will outweigh the disadvantages. For example, in a discussion of children finding unsuitable information on the Internet, Gates observes, "On balance, the advantages will greatly outweigh the problems."[23] On eliminating middlemen, Gates says, "But, as with so many changes, I think once we get used to it we'll wonder how we did without it."[24] We recognize the rhetoric of inevitability here.

We are in the middle of a difficult national debate about the privacy issues created by the collection of personal information for marketing purposes. There are increasingly effective computer-based means of capturing information about individual consumers, so ads can be directed at specific people whose preferences and precise demographics are known. The Internet will increase these opportunities manyfold. But Gates dispenses with huge privacy problems in a single sentence: "Preference data can be gathered and disseminated without violating anyone's privacy, because the interactive network will be able to use information about consumers to route advertising without revealing which specific households received it."[25]

This is technically possible, of course, but the economic incentives to sell preference data identifying specific individuals or households are enormous. Every time you visit a Web site, it is possible to record your visit and gather information useful to advertisers (especially if you are being paid to supply this information and view an ad, as Gates advocates). In the early 1990s, Lotus Development Corporation attempted to sell a CD-ROM with household data that identified specific households.

Lotus gave up this product after a great deal of negative publicity was generated (much of it carried on the Internet). But that doesn't mean that this can't and won't happen again.

Gates states that governments "should not attempt to design or dictate the nature of the information highway, because governments cannot outsmart or outmanage the competitive marketplace, particularly while there are still questions about customer preference and technological development."[26] In one breathtaking statement Gates defines the Internet as a consumer marketplace in which the outstanding issue is "customer preference," forgets that government-funded research programs *did* design the information highway,[27] and neglects to acknowledge that much "technological development" comes out of universities funded by governments, especially in the United States, Canada, Europe, Japan, and Australia.

In our view, the most egregious problem here is the attempt to convince people that the Internet is primarily a means of bringing together consumers and producers. While electronic commerce is a valid and productive use of the Internet, the Internet should be a platform for many different noncommercial uses as well. Today, the Internet is a diverse, protean, decentralized place, and there has been a remarkable creative response to its possibilities. Surprising and imaginative applications flourish in such an atmosphere. If the Internet remains open to creative and noncommercial experimentation, these applications will continue to emerge. Who would have thought five years ago that schoolchildren could control a powerful telescope, take their own pictures of the sky, and make scientific discoveries, as they have done in the Internet-accessible Hands-On Universe project?[28] These uses of the Internet enrich us all, and they do not fit within a vision of the Internet as "friction-free capitalism."

SUMMING UP

Throughout this book, we have portrayed information ecologies as local environments with personal meaning and real human connections. The vastness of the Internet seems to defy the essence of what a local habitation must be. But meaning and connection do not require physical proximity, and our technologies afford us new ways of becoming linked to one another. In information ecologies, we are bound to one another in specific relationships, but our interactions can be mediated through technology, including and perhaps especially the Internet.

Information ecologies can be grounded in an explicit ethos of universality, as well as emerging from geographically local concerns. Examples that come to mind are organizations devoted to serving refugees anywhere in the world, such as *Médicins sans Frontières* (Doctors without Borders) or the International Red Cross. Mother Teresa's Society of Missionaries operates in dozens of countries and takes its mission of service with it wherever it goes. The personality and instant recognizability of these organizations comes from their distinct and passionate involvement with a specific set of beliefs and goals. We expect that these organizations are precursors of many new international ecologies enabled by the Internet, which will derive their vitality and distinctiveness from shared values and purposes rather than shared geography.[29]

While writing this chapter, we ordered books through Amazon.com, sent email to Melinda Bier to get an update on her study (which we found posted on the Web), checked out sites to get information for this chapter and others, and used the Web in many of our other activities, including our own tiny information ecology that has developed during the two years we have worked on this book. Our hope is that the Internet will continue to flourish as the blooming buzzing confusion it is now—and even more so. It is an important source of diversity in an increasingly flat and packaged world. The Internet is a real challenge to Ellul, a significant counterexample to his predictions, so many of which seem to have come true.

If we nurture and defend local ecologies, the global network enabled by the Internet will avoid becoming a monocultural hegemony primarily devoted to commerce. The active participation of local nodes, mindful of their potential global connections, will lead to strength and vitality in a global network. We can take the exhortation of the environmental movement, "Think globally, act locally," a step further now. We can think and act both locally and globally. This is within the power of all of us with access to the Internet, whether we are heads of state or living on SSI in HUD housing.

13

Conclusion

This book was motivated by our research investigations into how people use technology in their local habitations and our many years of living in Silicon Valley. Our experiences leave us both excited and worried. We are excited because of what we have seen people do with technology in the information ecologies we described and because of our own personal experiences with technology.

We are worried because of the dynamics of *technique,* as articulated by Ellul, Winner, and others. The relentless buzz of technology as an engine of economic progress sounds in our ears every day. The marketing of technology emphasizes ever-increasing productivity—the ability to work faster, smarter, in more places at once, with people who are far away. "Anytime, anywhere," says the Silicon Valley mantra.

It sounds appealing to become a more advanced sort of person (worker, automobile driver, stock market investor, homeowner, parent), with the help of new, sophisticated tools. But how much faster and smarter *can* we work? Under what circumstances do we want to? Do we really want to be able to work anywhere, anytime, accessible to any and all, in a world of featureless temporal and spatial contours? What are the local consequences of working in these new ways? We know there must be important trade-offs involved, but the vision of limitless expansion and improvement through technology does not begin to identify them. To develop healthy information ecologies, we must explicitly address these trade-offs. We must address issues of "know-why" as well as "know-how."

What is at stake is the impact of our technologies on the quality of our lives. Because the rate of technical change is remarkably faster than in the past, we no longer have the luxury of slowly, organically evolving our practices to catch up with radically new technologies. We must deliberately consider our technology choices and changes to our practices in light of our values.

Jacques Ellul spoke of autonomous technology as though it were a personified force that somehow gave birth to itself. But technology comes from our own love of invention. We have seen this in our everyday lives around engineers. Much earlier, Lang symbolized this creativity in his character Rotwang. There are different versions of *Metropolis,* but in one of the most riveting, Rotwang's motivation for creating the robot is revealed: she is brought to life to recreate Rotwang's dead love. The robot is a symbol for our love of the technical, and it is meant to show how our own creativity is deeply implicated in this love.

Metropolis itself is a testament to technical creativity. Lang did what had never been done before. The film imaginatively portrays technologies that did not exist in 1926—instant electronic data access, doors controlled by sensors, video conferencing. Lang's art itself could only be fully realized through the use of what were then very advanced film and camera techniques, such as animation and morphing. The visual beauty of *Metropolis* is a masterful realization of Lang's deep connection to technology.

At the same time, Lang suffused *Metropolis* with an urgent sense of the troubling complexities and difficulties of technology. Technology regulates us: a giant clock face requires workers to struggle to maintain control of the machines by heaving a huge lever to precisely those points on the face that are lighted. Technology is always a step ahead of us: the hero tries in vain to prop open an automatically closing door with a large stick. Technology has the potential to completely overwhelm us: at the end of the film the machinery fails, and the city is immersed in rising waters flooding from an underground reservoir, endangering everyone— the "masters" of technology, as well as the workers and their children.

Ellul and other scholars are even more troubled by the direction of technological change than Lang was. While Lang clearly meant to harness technology for his own ends—including calling attention to its dangers— Ellul, Winner, Postman, and others are ready to slam on the brakes. We have learned from their brilliant analyses, but we also have felt the need to search for a more optimistic, empowering stance toward technological development.

In our research studies, we have been inspired by our encounters with people who use technology effectively and responsibly. In these settings, technology has been adapted to fit well with local practices and values. We found in Shakespeare's poetry a vivid expression of how creative invention becomes real and concrete when it is grounded in local meaning:

> And as imagination bodies forth
> The forms of things unknown, the poet's pen
> Turns them to shapes, and gives to airy nothing
> A local habitation and a name.

We have suggested that one way to enjoy the fruits of technology without succumbing to the very real perils of *technique* is to thoughtfully engage technology in our local habitations. The development of these local habitations, or information ecologies, is not the only solution to the problems of *technique*. It is the one we chose to write about because virtually everyone can influence their own ecologies.

We want to avoid being taken in by the rhetoric of inevitability. This rhetoric is powerful in part because it takes two seemingly opposite forms: the despair of dystopia and the don't-worry-be-happy optimism of technophilia. A key impediment to creating and nurturing robust information ecologies is believing (optimistically or pessimistically) that technical "progress" is ungovernable and inevitable. The most extreme but not uncommon manifestation of the rhetoric of inevitability is believing that any kind of technology is desirable as long as it can be reasonably engineered and manufactured.

We have adopted an ecology metaphor because it matches the dynamics we observed in the settings we studied. Information ecologies are composed of people, practices, values, and technologies. The characteristics of an information ecology share much with biological ecologies: diversity, locality, systemwide interrelationships, keystone species, and coevolution. What makes information ecologies different is the need to apply human values to the development of the practices and technologies within the ecology. We think of this as using technology with heart, after Lang's image of the human heart as the mediator between head and hands.

We have tried to add some new concepts and vocabulary to the conversations for action that we feel should accompany technological choices. We discussed the use of strategic questioning in information ecologies as a practical way to get started in the process of evaluating and shaping ecologies. The empirical chapters also contain a different kind of practical view of technology, because they show what people have done to use technology effectively in their ecologies. Each of our empirical studies highlighted some aspect of the ecological nature of technology use. In each, technology was used with heart, though in very different ways, according to the contours of the ecology.

In our study of reference librarians, we were struck by the "high-touch, high-tech" service librarians provided their clients. The latest technologies were in use, and they were used efficiently and effectively. But right alongside them was the enactment of the librarians' ethic of service. The librarians contributed their special human abilities of tact, diplomacy, judgment, and empathy. Their contributions turned the libraries we studied into places where clients felt comfortable and cared for at the same time they were receiving the benefits of the most advanced information technologies.

The library seems to us a place that exemplifies the importance of diversity in an information ecology. Here we saw a diversity of human and technical resources, working together smoothly rather than competing with one another. Librarians themselves are a keystone species, because without them and their clear value of service, there would be no library in its full sense. There would be a collection and some techniques for getting at the materials in the collection, but nothing more. As biologists have taught us, keystone species must be identified through extensive fieldwork—they are invisible unless you look and analyze carefully—and one of our goals here was to reveal some of the invisible work of a keystone species in the library ecology.

In Pueblo, we saw the dynamic of coevolving tools and practices at play. New tools were developed according to the local objectives and values of the school. Online rights and responsibilities centered around the needs and goals of a community of learners. Specific decisions about how to use technology, such as whether to permit the use of the gag command when students found themselves in conflict, were made according to the established classroom values of Longview Elementary. The value of community outreach was engaged as grays were brought in as new members of the Pueblo, providing an exciting new diversity to the makeup of the community.

We have made the point that tools and practices coevolve. At Longview, the use of the technology also changed some of the existing school practices. For example, the school negotiated with the district to be allowed to use students' work in Pueblo to meet district writing requirements.

Gardeners showed us how their unique mediation practices improve productivity and create a congenial atmosphere for work. Although Ellul would probably chide us for highlighting a practice that reinforces the dominion of technology, our observations suggest that skilled engineers and accountants enjoy their work, probably just as much as Ellul enjoyed his sociology. The idea that people have set things up so that they are supported in doing something useful and enjoyable seems to us, on balance, to be a good thing.

Gardeners themselves exhibit a kind of diversity as they use a number of different skills at once: technical, social, and organizational. They are tuned to the local pulse, responding with sensitivity and competence to the needs of their co-workers as these needs emerge within a local practice with its particular conditions and constraints. Gardeners also are tuned to the invisible interstices of their practice. As one gardener said, this is where the "pit crew" needs to be, if we can see past the flash and speed of the race car to the collaboration needed to win the race.

The Digital Photography class showed how a school's value of access equity was realized by focusing on essentials. One of these essentials was a sustained and purposeful emphasis on the underlying content of an activity rather than the tools employed in accomplishing it. The students and teacher of the photography class were there to learn an art form. Incidentally, they learned how to use some very fancy software tools—the same ones professional photographers use in their practice. But the tools never became the primary object of the exercise. As a result, the class appealed to a broad variety of students. Many of these students experienced in this class their first, vitally important successes in using computer-based technology, which can give them the confidence to find meaningful uses of computers in other contexts.

We also saw in the information ecology of the Digital Photo class that gardeners are just as valuable in school settings as they are in offices. The teacher of the photography class was an artist who initially approached technology with some trepidation. Her teaching skills were complemented by the contributions of the school's technology coordinator, who in classic gardener fashion was able to bridge the gaps between the advanced software and the local needs of the teaching staff.

The Internet is a vast playing field for diversity. It is a library, a marketplace, a post office, a retail shop, a broadcast service, and much more. As huge as it is, the Internet paradoxically underwrites diversity and the emergence of local information ecologies by providing a publishing forum and exchange mechanism for anyone with a computer and modem. The Internet is a giant network, but each node can be as small as one person or interest group. The Internet bypasses the bottleneck of the mass media (and government censorship in some cases) that so constrains expression. It is a medium in which we are seeing a return to handcrafted information—as though letter writing, pamphleteering, and afternoon watercolor painting are given new life. The letters are email messages, the pamphlets are posts to newsgroups, and the watercolors are digital photos now, but they are animated by the same spirit of personal expression.

The agenda for the Internet, as we see it, is to protect its healthy diversity and enlarge the pool of those who have access to it, through the appropriate installation of equipment and the provision of training and support. There are already viable models for making this happen, including the kind of "universal access" we provide for the telephone and the creation of telecenters, such as those planned for the Canadian Northwest. No one should be rendered invisible with respect to the Internet, unless they so choose. We must make sure we do not marginalize and exclude people who have not traditionally had access to information technology. As we have seen, the Internet has something to offer resistance fighters, elementary schoolchildren in Phoenix, stargazers, environmentalists, teenagers learning about their social history, low-income residents in HUD housing, purveyors of fine wines, and fish tank enthusiasts. We cannot think of a better recommendation for the Internet than its cast of characters.

In the hospital, we saw an example of how information ecologies can break down. High-quality patient care is the primary ethic of the hospital, but it is not the only one. In the teaching hospital, supporting the learning of physicians in training and fostering the collaborative workings of surgical teams were also important goals. When a new technology catered to the needs and interests of only one specialist at the expense of others, the ecology experienced tensions and dislocations.

We have to be willing to look at new technologies from the different perspectives of many people throughout the ecology, not just a few people who play the most visible and important roles. In addition, we learned from this example that it is far better to discuss and evaluate new technologies before they are introduced, to avoid having to repair bad feelings and lack of trust afterward. Even tools with unquestioned benefits can raise troubling issues. Through strategic questioning of possible consequences, diverse participants in an information ecology can improve a new technological design before it moves into active use.

Our way into the process of technological change is to adopt a stance of participation and engagement with technology, from the point of view of our local information ecologies. This is our sense of what it will take to penetrate the seemingly autonomous processes of *technique*. Resistance is sometimes a part of the strategy, but we believe it is a flawed approach if used by itself, because it disempowers. Using technology according to thoughtful values seems to us to be the most viable approach for the world we live in. We were amused, but also inspired, to read of Mother Teresa's fervent prayers and deep deliberations before she decided to add a single telephone to her mission house in Calcutta. After much thought, she felt the communication capability afforded by the telephone was important to the mission. But she also chose to carefully limit phone use and wait to see how it really felt to have the phone available.[1]

We have talked about the need to become more aware of our local habitations, to look at our own information ecologies with freshness and clarity of vision. In *Pilgrim at Tinker Creek,* Annie Dillard wrote about the difficulty of learning to see what was all around her:

> Seeing is of course very much a matter of verbalization. Unless I call my attention to what passes before my eyes, I simply won't see it. It is, as Ruskin says, "not merely unnoticed, but in the full, clear sense of the word, unseen." . . . If Tinker Mountain erupted, I'd be likely to notice. But if I want to notice the lesser cataclysms of valley life, I have to maintain in my head a running description of the present. . . . Otherwise, especially in a strange place, I'll never know what's happening.[2]

Changing technologies lend strangeness and adventure to many of the settings we inhabit. As Dillard says, we need to call our attention to what passes before our eyes. With a heightened awareness of our surroundings, we can know more about what is happening and understand better what actions we can take ourselves. We can also bring our values into our information ecologies, pay attention to the dynamics of coevolving tools and practices, and develop a sure sense of the importance of all of our contributions.

Notes

Preface

1. Quoted in Langdon Winner, *Autonomous Technology: Technics-out-of-Control as a Theme in Political Thought.* (Cambridge: MIT Press, 1977), 73.

Chapter 1: Rotwang the Inventor

1. The version of the film we use as the basis for this chapter is the 1989 video by Kino International Corporation.

Chapter 2: Framing Conversations about Technology

1. Arien Mack and Irvin Rock, *Inattentional Blindness* (Cambridge, MA: MIT Press, 1998).

2. The issue of invisible work is explored at length in *Computer-supported Cooperative Work—A Journal of Collaborative Computing* 8, nos. 1–2 (May 1998), with guest editors Bonnie Nardi and Yrjö Engeström.

3. Gordon Bell and James N. Gray, "The Revolution Yet to Happen," in *Beyond Calculation: The Next Fifty Years of Computing,* ed. Peter J. Denning and Robert M. Metcalfe (New York: Springer-Verlag, 1997).

4. See also Michael Dertouzos, *What Will Be: How the New World of Information Will Change Our Lives* (San Francisco: Harper San Francisco, 1997).

5. "Cloning procedure could bring unthinkable within reach," *San Jose Mercury News,* 24 February 1997.

6. "Cloning procedure could bring unthinkable within reach," *San Jose Mercury News,* 24 February 1997.

7. "Cloning procedure could bring unthinkable within reach," *San Jose Mercury News,* 24 February 1997.

8. Margaret R. McLean, "Just because we can, should we?" *San Jose Mercury News,* 18 January 1998.

9. Dystopic visions include Jerry Mander, *In the Absence of the Sacred* (San Francisco: Sierra Club Books, 1991); Sven Birkerts, *The Gutenberg Elegies: The Fate of Reading in an Electronic Age* (New York: Fawcett Books, 1995); and Neil Postman, *Technopoly* (New York: Vintage Books, 1993). Technophilia is well represented across the mass media in old-line publications such as *Time* and newer outlets such as *Wired.*

10. Nicholas Negroponte, *Being Digital* (New York: Knopf, 1995).

11. Clifford Stoll, *Silicon Snake Oil: Second Thoughts on the Information Highway* (New York: Doubleday, 1995).

12. Richard Sclove, *Democracy and Technology* (New York: Guilford Press, 1995).

Chapter 3: A Matter of Metaphor

1. See, for example, Jacques Ellul, *The Technological Society* (New York: Vintage Books, 1964); Ivan Illich, *Tools for Conviviality* (New York: Harper Colophon Books, 1980); Lewis Mumford, *Technics and Civilization* (New York: Harcourt, Brace & World, 1934); Postman, *Technopoly;* Langdon Winner, *Autonomous Technology: Technics-out-of-Control as a Theme in Political Thought* (Cambridge: MIT Press, 1977).

2. Donald Norman, *The Design of Everyday Things* (Garden City, NJ: Doubleday, 1990); Donald Norman, *Turn Signals Are the Facial Expressions of Automobiles* (Reading, MA: Addison-Wesley, 1993); Donald Norman, *Things That Make Us Smart* (Reading, MA: Addison-Wesley, 1994).

3. James J. Gibson, *The Ecological Approach to Visual Perception* (Hillsdale, NJ: Lawrence Erlbaum, 1986).

4. Bruno Latour, "Mixing Humans and Nonhumans Together: The Sociology of a Door-Closer," in *Ecologies of Knowledge,* ed. Susan Leigh Star (New York: State University of New York Press, 1995); Michel Callon, "The Dynamics of Technoeconomic Networks," in *Technological Change and Company Strategies,* ed. R. Coombs (London: Academic Press, 1992).

5. Illich, *Tools for Conviviality.*

6. Latour, "Mixing Humans and Nonhumans Together," 272.

7. Ellul, *The Technological Society,* xxv.

8. Ellul, *The Technological Society,* 6.

9. Ellul, *The Technological Society,* 6.

10. Ellul, *The Technological Society,* 14.

11. Ellul, *The Technological Society,* 134.

12. "Cloning ban recommended—Ethics panel says technique used by Scots is unsafe for humans," *San Jose Mercury News,* 8 June 1997.

13. Winner, *Autonomous Technology,* 73.

14. Winner, *Autonomous Technology,* 195.

15. Winner, *Autonomous Technology,* 85–86.

16. Winner, *Autonomous Technology,* 229.

17. Winner, *Autonomous Technology,* 229.

18. Winner, *Autonomous Technology,* 326.

19. A good introduction to the ideas of participatory design is found in Douglas Schuler and Aki Namioka's edited collection *Participatory Design: Principles and Practices* (Hillsdale, NJ: Lawrence Erlbaum Associates, 1993).

20. Winner, *Autonomous Technology,* 326.

21. Winner, *Autonomous Technology,* 330.

22. Jacques Ellul, "The Global Technological Systems and the Human Response" (paper presented via video recording at the Inaugural Meeting of the National Association for Science, Technology and Society, Arlington, Va., 5 February 1988).

23. Mark Philip, "Michel Foucault," in *The Return of Grand Theory in the Human Sciences,* ed. Quentin Skinner (Cambridge: Cambridge University Press, 1985).

24. There are complex arguments about the power of discourse that are beyond the scope of this book. Interested readers should refer to Michel Foucault, *Power/Knowledge: Selected Interviews and Other Writings 1972–1977* (New York: Pantheon Books, 1977).

Chapter 4: Information Ecologies

1. "High on tech—They're learning fast and loving it—Alameda company tries a maverick approach in hiring and training non-technical people for its customer service," *San Jose Mercury News,* 12 November 1997.

2. "What is your name?" Faust asked the Devil. To know a spirit's name was to put the spirit in the knower's power. (See translator's notes in Johann Wolfgang von Goethe, *Faust,* trans. Charles Passage [Indianapolis: Bobbs-Merrill, 1965, 49].)

Chapter 5: Values and Technology

1. A good introduction to issues of values and technology design is Batya Friedman's edited collection, *Human Values and the Design of Computer Technology* (Cambridge: Cambridge University Press, 1997).

2. Postman, *Technopoly.*

3. Postman, *Technopoly,* 70.

4. Jacques Ellul, "The Search for Ethics in a Technicist Society," in *Research in Philosophy and Technology,* ed. Frederick Ferr (Greenwich, CT: JAI Press, 1989), 26.

5. Jacques Ellul, *The Technological Society* (New York: Random House, 1964), 23–36.

Chapter 6: How to Evolve Information Ecologies

1. We found this poem circulating on the Internet.

2. "As chronic ills rise, toll on society grows," *San Jose Mercury News,* 13 November 1996.

3. Nobel prizes are awarded to individuals creating devices or devising tangible things like specific theories. Social studies of science show that the fruits of scientific and engineering efforts actually *are* team efforts but are socially constructed as the work of individual "geniuses"—clearly the point of view adopted by the Nobel committee. For example, see the work of Joan Fujimura in *Crafting Science: A Sociohistory of the Quest for the Genetics of Cancer* (Cambridge, MA: Harvard University Press, 1996).

4. Interview with Fritz Lang quoted in Frederick Ott, *The Films of Fritz Lang* (Secaucus, NJ: The Citadel Press, 1979).

5. Fran Peavey, *By Life's Grace: Musings on the Essence of Social Change* (Philadelphia: New Society Publishers, 1994).

Chapter 7: Librarians

1. Edward O. Wilson, *The Diversity of Life* (New York: W. W. Norton and Company, 1992).

2. Wilson, *The Diversity of Life,* 347–348.

3. Yvonne Baskin, "The Work of Nature," *Natural History* (February 1997).

4. The library ecology is also described in a research publication: Bonnie Nardi and Vicki O'Day, "Intelligent Agents: What We Learned at the Library," *Libri* 46 (1996). Vicki's study of librarians and library clients was done in collaboration with Robin Jeffries, and additional material not covered in this chapter is included in two research publications. One publication describes the evolving context of library clients' searches: Vicki O'Day and Robin Jeffries, "Orienteering in an Information Landscape: How Information Seekers Get From Here to There," in *Proceedings of the Conference on Human Factors in Computing Systems* (New York: ACM Press, 1993). Another publication describes how library clients share information within their local ecologies: Vicki O'Day and Robin Jeffries, "Infor-

mation Artisans: Patterns of Result Sharing by Information Searchers," in *Proceedings of the Conference on Organizational Computer Systems* (New York: ACM Press, 1993).

5. Sadly, in September 1997, in a budget-cutting move, the Apple Library was closed at the decision of the highest levels of management at Apple. The library's very fine collection was moved to the Stanford University Library. It was apparent from some comments made by management that they did not understand the nature of the work librarians were doing. For example, when the decision was being made about whether to close the Apple Library, a certain well-known, high-level executive was invited to the Apple Library to get a tour and talk to the head of the library about the services the library provided to employees. The librarian was explaining how the library was able to provide timely information to Apple engineers so they could do their jobs better. The executive stopped her and said, "They should already know all that, shouldn't they?" Another executive said they had made the decision to close the library because the presence of the library encouraged employees to waste time getting clever quotes for their slide presentations.

6. We should explain some of the standard practices in ethnographic writing that we have followed here. When we quote our informants or recount stories they have told us, we change names and other details to protect people's identities. We do not make changes to the spelling or grammar when we quote from email or other written material, because we feel it is important to preserve the original writer's own forms of expression. In excerpts from conversations in which we were participants, we use the tag "Interviewer" to refer to ourselves.

7. Terence Huwe, "Libraries and the Idea of Organization," *Advances in Librarianship* 21 (1997): 1–24.

Chapter 8: Wolf, Batgirl, and Starlight

1. The Pueblo research project had several sources of funding: a U.S. Department of Defense (ARPA) grant for education and technology under the CAETI (Computer-Aided Education and Training Initiative) program (with principal investigators Daniel Bobrow and Vijay Saraswat at Xerox PARC); the National Science Foundation (with principal investigators Daniel Bobrow at Xerox PARC and Billie Hughes and Jim Walters at Phoenix College), Phoenix College, and Xerox Corporation.

2. Pueblo has also been discussed in several research publications. The development of online learning activities is discussed in Vicki O'Day, Daniel Bobrow, Kimberly Bobrow, Mark Shirley, and Jim Walters, "Moving Practice: From Classrooms to MOO Rooms," *Computer Supported Cooperative Work— A Journal of Collaborative Computing* 7, nos. 1–2: 9–45. Special Issue on Interaction and Collaboration in MUDs, ed. Paul Dourish (Dordrecht, Netherlands: Kluwer Academic Publishers, 1998). The coevolution of technology and practice is discussed in Vicki O'Day, Daniel Bobrow, and Mark Shirley, "Network

Community Design: A Social-Technical Design Circle," *Computer Supported Cooperative Work—A Journal of Collaborative Computing* 7, nos. 3–4: 315–337. Special Issue on Participatory Design, ed. Jeanette Blomberg and Finn Kensing (Dordrecht, Netherlands: Kluwer Academic Publishers, 1998).

3. The technology in which Pueblo has been created is a kind of MUD (Multi-User Dungeon). Specifically, Pueblo is a MOO, which is one of many MUD variants. The acronym MOO stands for "MUD Object-Oriented"—the object-oriented part refers to the kind of programming language that is used to represent elements of the world. The MOO technology was originally developed by Pavel Curtis and Stephen White, and it is described in Pavel Curtis, "Mudding: Social Phenomena in Text-based Virtual Realities," *Intertrek* 3, no. 3 (1992): 26–34.

4. This earlier MUD was called MariMUSE.

5. Sherry Turkle, *Life on the Screen: Identity in the Age of the Internet* (New York: Simon and Schuster, 1996).

6. Donna Spano wrote these comments for the Pueblo project's final report to ARPA, one of the government agencies that funded the research.

7. These comments were also written for the Pueblo project's final report to ARPA.

8. Chip Morningstar and F. Randall Farmer, "The Lessons of Lucasfilm's Habitat," in *Cyberspace: First Steps,* ed. Michael Benedikt (Cambridge: MIT Press, 1994).

9. In some network communities, people form such close relationships that they decide to meet face to face for social gatherings. Howard Rheingold gives an example of this in his description of the Well (Whole Earth 'Lectronic Link) in his book *The Virtual Community: Homesteading on the Electronic Frontier* (Reading, MA: Addison-Wesley, 1993). More recently, we have seen this in SeniorNet, a network community for seniors that Vicki is currently studying. Many SeniorNet members feel that attending local "bashes" is an important complement to their online activity.

10. Teachers wrote these comments at the beginning of Pueblo summer camp, after completing their first full academic year of using Pueblo in their classrooms.

11. Jo Talazus wrote these comments for the Pueblo project's final report to ARPA.

Chapter 9: Cultivating Gardeners

1. These studies are reported in chapter six of Bonnie's book, *A Small Matter of Programming: Perspectives on End User Computing* (Cambridge: MIT Press, 1993).

Chapter 10: Digital Photography at Lincoln High School

1. Photoshop is a trademark of Adobe Systems.

2. The work reported in this chapter was conducted jointly with Brian Reilly. We produced a CD-ROM, "Digital Photography at Lincoln High School" that includes students' digital photos, video of classroom activities, audio and video interviews with students, the teacher, and the technology coordinator, as well as scanned images of the curricular materials used by Ms. Lynch. A QuickTime VR movie shows the classroom itself.

3. Not every school has a million dollars to devote to technology. Images taken with 35mm cameras or with inexpensive disposable cameras can be developed, scanned, and put on a Photo CD for a small fee by service bureaus. For even less money, other photo developing services will return prints along with a floppy disk containing digital copies of the pictures. With an image editing program such as Adobe Photoshop, you can crop, adjust, modify, and manipulate images for use in journalism or yearbook projects, Web pages, or any kind of student report or work that involves images. Digital cameras speed up the process, as they allow image capture and manipulation to take place at virtually the same time.

At Lincoln, Mr. Herlth started the Electronic Arts program, but we have also seen programs initiated by parent coalitions, principals, and teams of researchers and teachers. School-business partnerships are becoming more common and can be a source of funding and support.

4. For example, see Larry Cuban, "Why NetDay won't fix our schools," *San Jose Mercury News*, 3 March 1996. Cuban argued that the majority of future jobs will not require the use of computers and that the "substitution of virtual experiences for actual ones threatens to make the fake more compelling." The first assertion flies in the face of common sense, and the second is not borne out by our experiences or those in any of the classrooms we have studied. Creating digital photographs is no more "virtual" than writing an imaginary story. And it compares favorably with that most ubiquitous of virtual experiences, watching television.

5. Practical ideas for schools that take technology seriously include release time for teachers who plan innovative classroom uses of technology, participation in innovative teacher training programs, and programs to "cascade" expertise, so that a teacher who attends a training course is responsible for training a handful of teachers at her school upon her return, and they in turn train others.

Chapter 11: A Dysfunctional Ecology

1. This work was conducted jointly with Allan Kuchinsky, Steve Whittaker, Robert Leichner, and Heinrich Schwarz. It is reported in Bonnie Nardi, Allan Kuchinsky, Steve Whittaker, Robert Leichner, and Heinrich Schwarz, "Video-as-

Data: Technical and Social Aspects of a Collaborative Multimedia Application," *Computer Supported Cooperative Work—A Journal of Collaborative Computing* 4, no. 1 (1996): 73–100.

Chapter 12: Diversity on the Internet

1. Melinda Bier, Michael Gallo, Eddy Nuckols, Steven Sherblom, and Michael Pennick, "Personal Empowerment in the Study of Home Internet Use by Low-Income Families," *Journal of Research on Computing in Education* 30, no. 22 (1997): 106–119.

2. "Grant to bring Internet to housing project," *San Jose Mercury News,* 5 February 1998.

3. Reported in "Grant to bring Internet to housing project," *San Jose Mercury News,* 5 February 1998.

4. This statement was quoted in Melinda Bier, Michael Gallo, Eddy Nuckols, Steven Sherblom, and Michael Pennick, "Personal Empowerment in the Study of Home Internet Use by Low-Income Families," *Journal of Research on Computing in Education* 30, no. 22 (1997): 106–119.

5. Actually, teenagers may not be at such a disadvantage here. Vicki is currently working on an ethnographic study of an online community for seniors, and the discussion forum about men and women communicating is one of the few places where "flames"—agitated, even hostile Internet communications—occur. Maybe with decades of practice, Anthony and his friends will be better equipped to avoid flame wars on this topic by the time they are in their seventies and eighties.

6. *Cultural Survival Quarterly* (Winter 1998). Special Issue on the Internet and Indigenous Groups, with guest editor Steve Cisler.

7. The Oneida Indians had a Web page before the White House did. See Jean Polly, "Standing Stones in Cyberspace: The Oneida Indian Nation's Territory on the Web," *Cultural Survival Quarterly* (Winter 1998): 37–41.

8. Guillermo Delgado and Marc Becker, "Latin America: The Internet and Indigenous Texts," *Cultural Survival Quarterly* (Winter 1998): 23–28.

9. Barry Zellen, " 'Surf's up!' NWT's indigenous communities await a tidal wave of electronic information," *Cultural Survival Quarterly* (Winter 1998): 50–55.

10. Christina Fink, "Burma: Constructive Engagement in Cyberspace?" *Cultural Survival Quarterly* (Winter 1998): 29–33.

11. Fink, "Burma." Esther Dyson, an enthusiastic proponent of the Internet, says, "[The Internet] is a feeble tool for propaganda, but it's perfect for conspiracies." Her book, *Release 2.0, A Design for Living in the Digital Age* (New York: Broadway Books, 1997), is a good introduction to the Internet, covering a wide variety of topics including privacy, intellectual property, filtering, consumer protection, advertising, and the use of the Internet at work and in school. Dyson is also excited about the diversity promoted by the Internet and the possibilities for people to create as well as consume.

12. Mander, *In the Absence of the Sacred.*

13. Like many self-styled Luddites, Mander isn't completely against technology, he is just more comfortable with yesterday's technology. (For that matter, so were the original Luddites.) For example, Mander explains that he writes with an IBM Selectric typewriter rather than a computer. In some ways, this is fair enough—we all should think carefully about which new technologies we want to adopt and which we prefer to leave alone. But we also have to watch out for a tendency to hark back to the "paleoterrific."

14. Jerry Mander, *The Case Against the Global Economy* (San Francisco: Sierra Club Books, 1996).

15. Mander, *The Case Against the Global Economy.*

16. Bill Gates, *The Road Ahead* (New York: Viking, 1995).

17. Gates, *The Road Ahead,* 100.

18. Gates, *The Road Ahead,* 197.

19. Gates, *The Road Ahead,* 165.

20. Gates, *The Road Ahead,* 165.

21. Gates, *The Road Ahead,* 208.

22. Gates, *The Road Ahead,* 209.

23. Gates, *The Road Ahead,* 212.

24. Gates, *The Road Ahead,* 159.

25. Gates, *The Road Ahead,* 171–172.

26. Gates, *The Road Ahead,* 232.

27. Specifically, the Internet began with technology developed for the ARPANET project, funded by the Army's Defense Advanced Research Projects Agency.

28. The Hands-On Universe project searches for asteroids using images from the Berkeley Cosmology Project, a team of scientists searching for very distant supernovae. They use world-class telescopes such as the Cerro Tollolo International Observatory in Chile to search for supernovae near the edge of the visible universe. The scientists generously share their data with Hands-On Universe classes so that students can search for very faint asteroids in the same regions of the sky. To date, five unknown asteroids have been recorded by Hands-On Universe students.

29. *Médicins sans Frontières* and the Red Cross have Web sites. Mother Teresa has many sites *about* her, though she departed this world without her own home page.

Chapter 13: Conclusion

1. Raghu Rai and Navin Chawla, *Mother Teresa: Faith and Compassion, the Life and Work of Mother Teresa* (Dorset, UK: Element Books, 1996).

2. Annie Dillard, *Pilgrim at Tinker Creek* (New York: Bantam Books, 1974), 32.

Index